Contents

HEALTH AND SAFETY
FOR SMALL BUSINESSES

HEALTH AND SAFETY
FOR
SMALL BUSINESSES

Tom O'Reilly

2000

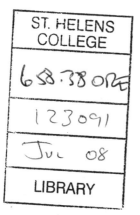
First published in 2008 by Management Books 2000 Ltd
Forge House, Limes Road
Kemble, Cirencester
Gloucestershire, GL7 6AD, UK
Tel: 0044 (0) 1285 771441
Fax: 0044 (0) 1285 771055
Email: info@mb2000.com
Web: www.mb2000.com

British Library Cataloguing in Publication Data is available
Produced by Digital Book Print Ltd
ISBN 9781852525743

Preface

While entrepreneurs and business owner/managers are struggling to design and produce competitive successful products, to create a market for those products, and to grind out a profit in these ever more challenging times, it is all too easy to ignore the more mundane aspects of health and safety. But the reality is that much of the current health and safety legislation is essential to protect your most valuable asset – your staff. The trick is to establish practical and pragmatic approaches to the requirements which meet the reasonable objectives and avoid getting bogged down in bureaucratic detail. If you plan properly from the start, Health and Safety does not need to be exorbitantly expensive; it can be much more costly if you don't observe the legal requirements and either an accident happens or the dreaded inspectors catch up with you...

There is a whole expensive library of legal textbooks under the general heading of health and safety which will provide chapter and verse on all the detailed requirements of current health and safety legislation. There is also an army of health and safety consultants ready and all too willing to provide you with expensive advice. However, for the small business owner there is rarely time to go through all the minutiae. This simple book will help to guide you inexpensively through the legal maze, and provide some quick and practical solutions to the majority of the requirements. It is directed at small businesses, defined under European legislation as a company having less than 49 employees, of which there are 3-4 million in the UK.

Health and safety is important and legislation is constantly changing. It is just a pity that the "law is an ass" syndrome is so often encountered in the field of health and safety. Ever more asinine laws are generated on a daily basis from the European Parliament. The present Government apparently accepts European diktats without demur and continues to expand its public sector with quangos and over-zealous inspectorates determined to carry out every legislative whim of the European Council of Ministers in Brussels. As an employer, you mustn't let this detract from the importance of health and safety in your business arrangements. Only a small proportion of the welter of health and safety law will apply to you, so don't despair. It is always essential to obtain the most up-to-date information before committing yourself to unnecessary expense.

The broad aim of this book is to provide you with the essential basic information and some practical guidelines which will help you to steer clear of the "health and safety storm-troopers" and to comply with those parts of the health and safety regulations which may apply to your enterprise.

Tom O'Reilly
June 2008

Introduction

Early health and safety laws were reactive in response to problems in particular industries, or particular premises – factories, mines, shops, offices and railway premises. This left large sections of the working population unaffected and unprotected by any safety regulations. To remedy this, the *Health and Safety at Work etc Act 1974* (HSWA) was introduced to cover employers and employees, rather than premises or particular industries. This brought millions of new workers under the umbrella of safety legislation for the first time. The Act established a Health and Safety Commission (HSC) empowered to deal with all health and safety matters including research, arrangements for training, advice to government departments and employers, and submitting proposals for amending old laws and introducing new legislation and codes of practice. A second body, the Health and Safety Executive (HSE), is the executive arm of the HSC and it is this organisation which you will come across more often in connection with health and safety matters.

The Act starts by dealing with an employer's duty towards his employees. It then goes on to deal with the employer's duty towards others who might be affected by the business – contractors, visitors and members of the public. This duty also covers the self-employed. Next on the list are designers, manufacturers, suppliers and importers who have a duty to make sure that any articles or substances they produce or supply are safe to use, properly tested, examined and accompanied by the relevant safety instructions. Next come employees themselves as they too have certain obligations under the Act – they must take care of their own and others' health and safety at work, co-operate with their employer, and not interfere recklessly with, or misuse, any health and safety equipment provided by their employer.

Complying with, and implementing, health and safety measures can involve considerable (and often unexpected) expense. These are costs which are often overlooked in preparing the start-up business plan, and consequently may not be budgeted for. The bad news is that since 1974, health and safety legislation has grown and the HSWA is merely the "nursery slopes" of an ever-increasing mountain of paper. Moreover, membership of the EU has meant that the member states have had to assimilate Brussels-originated directives into their own law. This has resulted in a proliferation of European-generated regulations and red tape which is strangling small businesses. It has been said of European health and safety directives that the Germans decree them, the French and Italians ignore them, the Spanish are blissfully unaware of them, and the British obey them. You only need to holiday on the Continent to see how religiously our European neighbours abide by the same health and safety laws! Now that the EU has expanded to include 27 different countries, local interpretation of law will be even more diverse.

The red tape is not just limited to health and safety legislation. Bureaucracy is having a field day with small businesses, with a whole raft of restrictive legislation, some useful and some less so – covering such areas as anti-discrimination law, working hours, trade union recognition, maternity and paternity leave, minimum wages and so on. However, health and safety is the most complicated, and potentially the most damaging area of state intervention. With so many bewildering rules and regulations it is difficult to know where to begin but, if you are occupying premises and employing people, there are some practical things you should deal with first.

Let's examine some of the important health and safety priorities in greater detail. Take a look at the following questions. These are basic health and safety considerations relevant to many businesses.

- Have you registered with the HSE or Local Authority?
- Have you taken out insurance?
- Do you need a health and safety policy?
- Do you need an Accident Book?
- What about fire precautions? Are employees trained in fire procedures, prevention and evacuation? How often do they participate in fire-evacuation training? Who co-ordinates your evacuation procedures?
- Are employees trained in first aid?

- What are risk assessments?
- What is Personal Protective Equipment?
- Is your office/shop a workplace?
- Is your work equipment safe to use?
- Do you know how to do a COSHH assessment?
- What is Display Screen Equipment?
- What about manual handling?
- What information must I give my staff?
- What about training?
- Are any of your staff disabled?
- Do you understand the anti-discrimination rules?
- How are visitors and contractors controlled when on your premises?
- What must employees do?
- How do I manage health and safety?
- What happens if I breach health and safety law?
- What do I do about smoking?
- What do I do about waste and pollution?

You can see from the list (and that's not the end of it!) that there is enough regulation and red tape to discourage all but the most determined. However, by adopting a common-sense and practical approach you will find most of the regulations easy to adapt to your needs – and you can take some comfort from the fact that not all of the legislation will apply to your particular business. The final chapter of this book contains a potted history of health and safety providing some historical perspective and moral justification for the law as we know it today. You should make a start with the practical things you need to do and the first step is to register your business.

1

Registration

(Note: Do not confuse health and safety registration with registering a limited company with Companies House. If you are forming a limited company you can obtain advice on how to register from www.smallbusiness.co.uk. Health and safety registration is a separate process, and applies to all employers, whether registered as limited companies or not.)

Background

In the early 1800s, a Royal Commission recommended the appointment of inspectors to inspect some factories. The inspectors were given the right to enter factories and to question workers. This laid the foundations for the current regulatory system, as defined in the *Health and Safety at Work etc Act 1974*. The HSWA established a Health and Safety Commission (HSC) and a Health and Safety Executive (HSE) with the following principal roles and responsibilities. The Executive consists of three persons appointed by the HSC with the approval of the Secretary of State. Its main function is to carry out and give effect to any of the HSC directions and to make adequate arrangements for the enforcement of health and safety regulations. Since 1974, inspectors have been authorised by both the Health and Safety Executive and local authorities. Inspection has been divided between them depending on the nature of the work or workplace.

The appointment of inspectors is made in writing specifying their powers. They may only exercise their powers within the field of responsibility of the authority which appointed them. They must produce, on demand, their instrument of appointment. Inspectors may enter any premises in order to exercise their powers; be accompanied by a police officer if there is reason to believe serious obstruction might be met in the execution of duty; make any necessary examinations and investigations; direct that those premises or part of them or anything in them, e.g. machinery, shall be left undisturbed for the purpose of investigation or examination; take photographs and such measurements as are deemed necessary; take samples, e.g. dust, of any articles or substances found in any premises (where a sample has been taken, the inspector shall give a portion of it to a responsible person); dismantle any article or substance which is likely to cause danger to health and safety or subject them to any process or test; retain any article or substance to ensure it is available for use as evidence in any court proceedings; require any person (whom they have reason to believe may be able to give relevant information about the investigation) to answer questions and to sign a declaration as to truth of their answers; require the production of, inspect, and take copies of, or of any entry in, any books or documents which are required by regulations (having taken possession of any article or substance, they must leave a notice at the premises with a responsible person stating that he they have taken possession – if that is not practicable, the notice must be fixed in a conspicuous place); issue improvement or prohibition notices.

Both HSE and local authority inspectors have the authority to prosecute in court for breaches of health and safety law. You are obliged therefore to register with your enforcing authority. This doesn't mean that if you don't register you can't be inspected. As mentioned above, inspectors have wide-ranging powers to enter any premises to ensure that businesses are complying with health and safety regulations. Whether in fact you receive a visit by an inspector will often depend on the nature of your business and your accident record.

Do you need to register?

Almost all businesses need to be registered with either HSE or the Local Authority (see below to determine which). However, there are a few exceptions:

- Places where only self-employed people work.
- Businesses where the only persons employed are close relatives of the employer, i.e. husband, wife, parent, grandparent, son, daughter, grandchild, brother or sister of the persons by whom they are employed.
- Outworkers' dwellings (if they are the only persons employed there).
- Premises where the sum of hours normally worked each week by all employees is not more than 21.
- Premises occupied solely by members of the armed forces or of the armed forces of certain other countries.
- Premises used for selling fish wholesale in docks.
- Parts of mines below ground.

If you think you may already be registered, you should check by contacting the relevant authority. (Note that just because you are paying business rates does not necessarily mean that you are registered for health and safety purposes with the local authority.)

In all cases registration should occur prior to occupation of the business premises.

Which authority do you register with?

That depends on your type of business. The HSE deals with, amongst others, offices and shops in factories and other places covered by the *Factories Act 1961*; premises owned or occupied by the Crown; county councils; and local police and fire authorities. Factories, mines and quarries, building sites, schools, hospitals and the larger organisations such as railways, fairgrounds, chemical plants, nuclear and offshore installations such as oil rigs are also dealt with by the HSE.

Local authorities deal with offices and shops, warehouses, hotels and other catering establishments open to the public, sports and leisure complexes, and fuel storage premises. Office premises means a building, or part of it, which is used for office purposes, and 'office purposes' includes administration, clerical work, book-keeping, filing, typing, drawing, etc.

You can see from this that there is some overlapping in respect of offices. If you are not sure which authority to register with you can telephone the HSE or your local authority for advice.

Registering with the HSE

If you are registering with the HSE you need to obtain a form F9 (shown on page 18) which is a *notice of factory occupation.*

For further information, or to download a form for online completion, visit www. hse.gov.uk.

Form F9

Factories Act 1961
Health and Safety at Work etc Act 1974

HSE
Health & Safety
Executive

Notice of occupation

FOR OFFICIAL USE ONLY
Complete before despatch

For attention of:

Section 137 of the Factories Act 1961 requires every person intending to occupy or use any premises as a factory to serve on HM Inspector of Factories a written notice giving the particulars specified below. At least one months notice must be given before the occupation or use begins.

If a person takes over from another without changing the nature of the work, previous notice of one month need not be given. However the notice must be served as soon as practicable and in any event within one month of his taking over. In other cases the occupation or use may begin less than one month after the notice if the Inspector gives written permission.

The Data Protection Act 1998 requires the Health and safety Executive (HSE) to inform you that this form may include information about you (this is called 'personal data' in the Act) and if we are a 'data controller' for the purposes of this Act. HSE will process the data for health, safety and environmental purposes. HSE may disclose these data to any person or organisation for the purposes for which it was collected or where the Act allows disclosure. As data subject, you have the right to ask for a copy of the data and to ask for any inaccurate data to be corrected.

When completed this form should be sent to the local Health and Safety Executive area office.

Name of occupier or title of the firm

Full postal address of the factory (please include the Postcode)

Registered office, if any

Nature of the work to be carried on

Is mechanical power to be used for any purpose other than the heating, ventilating, or lighting the workroom or other parts of the factory? Yes ☐ No ☐

If 'YES', state its nature (eg electric, steam, gas or oil)

If 'NO' and mechanical power is subsequently use for any purpose other than those mentioned, the occupier should notify an Inspector of the HSE at the area office no less than one month before the date on which it is first used. The nature of the mechanical power should also be given - section 137(2) of the Factories Act 1961.

Name of District or Island Council within whose area the factory is situated.

I hereby give notice that (Please tick appropriate box)
I intend to occupy ☐ a factory as detailed above
I am occupying ☐ a factory as detailed above

Name

Present address (including postcode)

Telephone number

Signature   Date
F9 (01.02)

18

Registering with your local authority

If you are registering with your local authority, you need form OSR1 (shown below) – a *notice of employment of persons in office, shop or certain railway premises*. (Visit www. hse.gov.uk to download a copy of the form for online completion.)

Form OSR1

The Data Protection Act 1998 requires the Health and Safety Executive (HSE) to inform you that this form may include information about you (this is called 'personal data' in the Act) and that we are a 'data controller' for the purposes of this Act. HSE will process the data for health, safety and environmental purposes. HSE may disclose this data to any person or organisation for the purposes for which it was collected or where the Act allows disclosure. As data subject, you have the right to ask for a copy of the data and to ask for any inaccurate data to be corrected.

HSE
Health & Safety
Executive

Office Shops and Railway Premises Act 1963

Notice in form prescribed by the Secretary of State for Employment, of employment of persons in office or shop premises.

Click here for guidance

If you intend to employ any person or persons to work in shop or office premises you are required by Section 49 of the Offices, Shops and Railway Premises Act 1963 and the Notification of Employment of Persons Order 1964, to complete this form and send it to the appropriate authority. Please read the explanatory notes by clicking the 'Click here for guidance' link above before completing this form.

The appropriate authority will send a copy of the form to the fire authority for your area. (You may need a fire certificate -see note 9 in the guidance notes)

A separate form should be completed for each set of premises with a different postal address. Where several occupiers have premises at the same address, each occupier should complete a form in respect of his premises.

When completing this form please ensure you select the relevant checkbox for Part I or Part II, whichever is applicable.

Part I

☐ Notice is hereby given that on the *(insert date)* [_____], the employer specified in Part III of this notice, will begin to employ persons to work in the premises described therein.

Part II

☐ Notice is hereby given that the employer specified in part III of this notice is employing persons to work in the premises described therein.

Part III

1 (a) Name of employer

(b) Trading name, if any

2 (a) Postal address of the premises

(b) Telephone number

3 Nature of business

4 How many persons are or will be employed by the employer in office or shop premises at the above address in the following types of workplace? (see notes 3-7)

(a) Office

(b) Shop (retail)

(c) Wholesale department or warehouse

(d) Catering establishment open to the public

(e) Staff canteen

(f) Fuel storage depot

Total 0

Of the total how many are females?

5 How many of the total are or will be employed on floors other than the ground floor?

6 Of the total stated in reply to question 4, are any (or will any be) housed in separate buildings? Yes ☐ No ☐

7 Is the employer the owner of the building(s) (or part of the building(s)) containing the premises? Yes ☐ No ☐

8 If not, state the name and address of the owner(s) or person(s) to whom rent is paid

Signature of employer or person authorised to sign on his behalf [_____] Date [_____]

OSR1 (11.92)

19

2

Insurance

Liability and insurance

The first thing the *Health and Safety at Work etc Act 1974* did was to impose a duty on employers to protect, *so far as is reasonably practicable*, the health, safety and welfare at work of all his employees. But what does "reasonably practicable" mean? Let's try to translate some of the legislator's gobbledegook into understandable English. "So far as is reasonably practicable" is an important phrase in the world of health and safety. The phrase has been examined in our Courts a number of times and the legal definitions emerging are expressed in the usual "legalese" terminology. In plain English, it means that in deciding to carry out any work involving a risk of injury or damage which is qualified by the phrase, you must weigh the degree of risk involved against the cost of avoiding or reducing the risk. If the risk is small and the cost of avoiding the risk is very high then there is no need to implement the measures to reduce the risk. It is important to understand that cost here means time or trouble and not simply money. In the principal case, a miner was killed by a fall of rock whilst walking along an underground roadway (*Edwards v National Coal Board 1949*). The mine owners were obliged to ensure that the roof and sides of underground roads should be made safe. However, they had a defence to a charge of negligence if they could prove that it was not *reasonably practicable* to guarantee the safety of every "travelling road and workplace". You may feel that to make every roadway in a coalmine safe would have been a tall order but the judges in the case held that this was in fact the standard required. It emphasises the standard you will have to maintain in your presumably less dangerous working environment!

Health, safety and welfare, as mentioned, cover everything in the workplace:

- the provision and maintenance of safe plant and systems of work;
- safety in the use, handling, storage and transport of articles and substances;
- the provision of information, instruction, training and supervision;
- the maintenance of the workplace in a safe condition with safe access and egress;
- a safe working environment and adequate welfare facilities; and
- a written safety policy.

Lots of things to go wrong – and ultimate responsibility and liability for all of these matters rests fairly and squarely on you, the employer.

You may delegate arrangements for managing health and safety but your actual responsibility (and hence your liability) for health and safety itself cannot be delegated.

What is your 'liability'?

Liability means being subject to a legal obligation which can be criminal or civil according to whether it is enforced in a criminal or a civil court. **Vicarious liability** needs some explanation as it crops up continually in health and safety law. "Vicarious" means "experienced at second hand" through the actions or experiences of a substituted or delegated third party. In the context of liability this means that you are liable for the actions of others acting on your behalf (as your substitute). Under employment law, you, as an employer, are liable for the wrongful acts or omissions of your employees provided that the acts or omissions occur within the scope of their employment, i.e. at work or arising from work. In a case where a third party is injured by an action or omission of one of your employees, the claimant (injured party) can sue both the employee personally and you, the employer, as you are vicariously liable. In practical terms, the employer is more often sued as he is more likely to be in a position to pay damages than an employee.

Liability comes in several forms:

- **Criminal liability** is incurred when a breach of the criminal law takes place. The person responsible for the breach is liable and must suffer the consequences.

- **Civil liability** arises from an act or omission (a "tort") which gives one individual or body the right to present a legal claim against another for restitution. A tort in English law is a breach of a duty imposed by law. It applies in a number of areas – nuisance and trespass being particularly common – *but for the purposes of health and safety the most important is the tort of **negligence**.*

- **Absolute liability** is often referred to as strict liability and applies equally in criminal and civil law. It arises when an offence governed by absolute liability is committed. Sections of law are often written in absolute terms where duties and obligations are qualified by the word *"shall"*. Where this occurs, the offender has little room to manoeuvre. An example of absolute employer's liability is as follows:

  "Every employer shall ensure that work equipment is so constructed or adapted as to be suitable for the purpose for which it is to be used or provided".

  If you, as an employer have failed to ensure that the equipment is so constructed or adapted, you are obviously in breach of the law and therefore liable. In such cases, whether you had good reason to commit or did not intend to commit the offence is immaterial.

- **Corporate liability** is where a company or corporation (rather than an individual) has liability. A corporation has a separate legal personality and can as a result be guilty of a criminal offence. However, Government departments escaped prosecution because of Crown immunity. The problem in prosecution generally lies in the fact that to prove the guilt of a criminal offence the prosecution must prove that the offender had a "guilty mind or guilty intent". It is difficult in the case of a board of directors to identify which individuals are, in fact, the board's mind and will. To deal with this anomaly a new law, the *Corporate Manslaughter and Corporate Homicide Act 2007*, was passed in July 2007. This law removes the "guilty mind" requirement. Now, an organisation which causes a person's death because its activities have been so badly mismanaged (amounting to a gross breach of the duty of care) is guilty of corporate manslaughter. "Gross" is interpreted as meaning if the alleged conduct falls far below what can be reasonably expected of the organisation in the circumstances.

- **Employers' liability**. This is the important one as far as you are concerned. That things go wrong in the workplace is inevitable, and once you start employing people you become responsible for their health and safety whilst at work. Bodily harm (including mental or psychiatric illness) or disease sustained by any of your employees arising out of and in the course of your business is your responsibility and you must insure against liability for it.

Who is an employer?

Put simply, an employer is a person who employs others under a contract of employment.

Who is an employee?

Again, the simple answer is an employee is one who works for an employer under a contract of employment. Simplicity itself until you consider the position of contractors and casual workers. And what about **relatives** who work for you? Here, there is some guidance as close relatives – spouses, fathers, mothers, grandparents, etc. as mentioned above – are exempted from the need for liability insurance unless you are operating as a limited company.

Liabilities under health and safety law

If a prosecution results from a breach of health and safety law you may be held liable. Similarly if an employee is injured as a result of an accident at work, you may be sued for damages in the civil courts. Even more worrying is the fact that you can be prosecuted in the criminal court for a breach of health and safety law and sued in the civil court for negligence arising from the same set of circumstances.

Employers' liability insurance

This is covered by the *Employers' Liability (Compulsory Insurance) Act 1969*. Amongst other provisions, the sum insured must be a legal minimum of £5 million (although standard practice by most insurers is to provide cover for at least £10 million) to cover claims for personal injury or disease sustained by an employee as a result of his employment. You must then display your **Employer's Liability Insurance Certificate** in a prominent place in your premises. If you have more than one place of business you must display it in each location where your employees work. Failure to display a current certificate can carry a fine of up to £1,000 and companies without the necessary insurance face fines of up to £2,500 per day.

Because of the latency of some diseases and medical conditions, e.g. mesothelioma and industrial deafness, employers must retain copies of out of date policies for 40 years.

In 2004, the Government had a rush of blood to the head and recognized, in a blinding flash of the obvious, that a sole trader was unlikely to sue himself for injuries sustained at work and announced that, as from 2005, *companies which employ only the owner would be exempt from buying employers' liability insurance.*

Obviously, insurance liability goes further than just employers' liability and insurance doesn't come cheap. A small manufacturing business with a dozen or so employees would probably buy insurance cover for employers' liability, public liability, product liability, buildings, motor, stock and plant at an annual cost of some £4-5,000.

Occupiers' liability

The *Occupiers' Liability Act 1957* deals with your duty towards lawful visitors to your premises or place of work. As the occupier of premises, you owe what is known as a common duty of care to any visitors. Simply put, this means that you must take reasonable steps to ensure that any *lawful* visitors to your premises are safe from harm. You are liable for any injury sustained by a visitor to your premises as a result of your negligence. It is surely common sense to inform visitors of any known structural defects in your building and to point out any obvious dangers in connection with any work being carried out on the premises.

A simple example of the exercising of a duty of care is where you see on the supermarket floor a warning notice where something has been spilled and then mopped up.

Under a second *Occupiers' Liability Act 1984*, you also owe a duty of care to a *trespasser* on your premises. This is one of those almost unbelievable "law is an ass" bits of nonsense until you look more deeply into the reasons for it. For example, not all trespassers are wilfully on your premises. What about a child or a mentally-handicapped person? Would you reasonably expect them to be as careful or aware of potential danger as an adult? If you know that there are dangers present in your workplace, which they would fail to recognise as such, you owe them a greater duty of care and are reasonably expected to offer them some enhanced degree of protection.

To protect yourself against claims under occupiers' liability it is sensible to take out **public liability insurance**.

Advice on all insurance matters can be obtained from any reputable business and commercial insurance broker.

3

Health and Safety Policy

The need for a written health and safety policy is essential as it is the key document setting out how your business is to fulfil its obligations under the Act. Most policy statements simply reproduce the HSWA in non-parliamentary language, and you can follow suit by:

- identifying and controlling any health and safety risks arising from your business
- consulting with your employees on health and safety matters
- providing and maintaining safe plant and equipment
- ensuring safe handling and use of substances
- providing information, instruction and supervision
- employing competent staff and giving them proper training
- preventing accidents
- maintaining safe and healthy working conditions.

Your policy document should set out the organisation and arrangements for implementing your policy. You must review and revise it when necessary and at regular intervals. Having taken the time and trouble to produce it, don't file it away – you must bring it to the notice of all your employees. *Where you have five or more employees your policy must be in writing.* Where you have fewer than five employees, it may be in verbal form. This is another piece of legalistic nonsense. Proving to a court or tribunal that you have a verbal policy is going to be argumentative. It's a bit like Sam Goldwyn's famous saying – "a verbal contract ain't worth the paper it's written on".

In today's litigious society it is advisable to have a written policy regardless of the number of staff you employ.

After all, preparing a policy statement is not difficult. For a small business it could be confined to a single sheet of paper.

Apart from demonstrating your compliance with the law, your policy informs your employees and contractors about your health and safety arrangements, is helpful in obtaining contracts and signifies your professionalism to insurance companies and potential clients. On the question of obtaining contracts, you will be asked to send evidence of your health and safety commitment along with your tendering documentation – what better evidence than your safety policy?

Publishing the policy

There are various ways of bringing the policy to your employees' notice. Posting it on your staff notice board may well be sufficient to comply with the requirement. In other situations, with a larger workforce, it may be necessary to issue staff with a personal copy. Probably the safest means is to give each employee a copy contained in a staff handbook. In this computer age it is a simple matter to put your policy on the company computer system. Any revisions of the policy can then be dealt with easily. What you must do is to provide a document that deals with general policy, organisation, and the arrangements for bringing your general policy into effect.

There can be other communication problems. What if you have members of the workforce who do not understand English? They are entitled to the same protection in law as those who can understand the language. You obviously need to produce a statement in a language they can understand.

Writing the policy

It is neither rocket science nor nuclear physics so don't employ an expensive consultant to write it for you!

The policy document has three parts: the statement of intent, the organisation, and the arrangements.

The statement of intent

This outlines your commitment to health and safety and defines your objectives in complying with the law. This commitment should embrace everyone – all company directors and managers and the rest of your staff. The objectives would include, amongst other things, the reduction of accidents, consultation with employees, and putting health and safety on a par with production, finance, quality control, etc.

The organisation

The organisational part of the policy document is the answer to the question, "Who is responsible for health and safety?" It should describe who is going to be responsible for achieving your health and safety objectives on a day-to-day basis. You need to designate a competent person or persons to do this.

Do it by job title rather than by individual's names to avoid having to change the document whenever staff leave or are promoted, e.g.

- *Managing Director:* providing the necessary resources to implement H&S Policy; reviewing performance; reviewing policy annually.
- *Manager:* ensuring that premises and equipment are properly maintained; ensuring that safe systems of work are maintained; ensuring that adequate training is provided; chairing health and safety committees; ensuring that risk assessments of items which affect the safety and health of employees, including Manual Handling, Display Screen Equipment (DSE), Personal Protective Equipment (PPE), Provision and Use of Work Equipment (PUWER), and Fire safety are carried out.
- *Supervisor:* ensuring that employees under his control are complying with company health and safety policy in relation to PPE, Control of Substances Hazardous to Health (COSHH), and accident procedures; risk assessments.

- *Health and Safety Manager:* liaising with HSE, local authority, Fire Officer, etc.; investigating incidents/accidents, and reporting under the *Reporting of Injuries Diseases and Dangerous Occurrences Regulations* (RIDDOR); identifying training needs; providing or organising health and safety training; carrying out inspections and audits; co-ordinating risk assessments; testing evacuation procedures; producing the company 'Management of Health and Safety Action Plan' and preventive and protective measures based on risk assessments; acting as secretary to health and safety committees.
- *Employee:* complying with Company health and safety procedures; acting responsibly at all times; reporting defects, incidents and accidents; co-operating at all times with management on health and safety matters.

Arrangements

The document must specify the arrangements for implementing your policy. They will cover work practices such as risk assessment, consultation with employees, maintenance of safe plant and equipment, handling and use of substances, information, training and supervision, first aid and accidents, and fire and emergency evacuation procedures. More specifically they should include:

- medical surveillance as required for those likely to be affected by work activities, for example crane and fork-lift truck drivers, welders, and persons exposed to soldering fumes;
- the provision of easily understood information on risks to health and safety;
- preventive and protective measures;
- the name of the person(s) in charge of first aid, fire and emergency evacuation;
- the name of any safety co-ordinator for multiple occupancy of a site and details of any safety procedures;
- details of risk involved on "host" premises, i.e. rules for sub-contractors;
- the training to be given on recruitment, or changes in systems of work, changed technology, or new risk;

- the need for employees to inform you of dangerous work situations or defects in the preventive or protective measures (this is usually covered by the use of a **"Defects Book"**).

All this is a rather large mouthful for a safety policy document. It is a good idea to adopt a broad-brush approach in stating your arrangements in the policy and transferring the detail to a separate company manual. In that way, the manual can be edited to deal with minor changes and changes in personnel – leaving the policy document intact.

Monitoring and reviewing the policy

You must monitor and review your safety policy periodically to deal with the introduction of new technology, new systems of work and changes in health and safety legislation. Your current policy must be dated and signed by you as managing director, as you are primarily responsible for all health and safety matters, and so that it can be recognised as an authoritative document.

Specimen policy

Take a look at the following specimen policy which can easily be modified to suit your particular business.

Specimen Company Health and Safety Policy

J. Bloggs Ltd is fully committed to meeting its obligations under the Health and Safety at Work etc Act 1974, the Management of Health and Safety at Work Regulations 1999, the Regulatory Reform (Fire Safety) Order 2005, and associated legislation. To achieve those objectives it has appointed designated members of staff to be responsible for managing Company health and safety in order to keep workplace health, safety, welfare and new legislation under constant review in order to ensure ongoing compliance with the law.

The main responsibility for all health and safety matters remains with the Managing Director and Board of Directors who are bound by any of their acts and/or omissions which give rise to legal liability – provided that such acts or omissions arise out of and in the course of Company business.

To comply with both statutory and common law duties, the Company has arranged insurance against liability for death, injury and/or disease suffered by any of its employees arising out of and in the course of employment if caused by negligence and/or breach of statutory duty on the part of the Company.

For their part, employees agree contractually to comply with their individual duties under all relevant health and safety regulations and will co-operate with the MD to enable him to carry out his duties and obligations under the Health and Safety at Work etc Act 1974 and associated legislation. Failure to comply on the part of any employee may lead to dismissal. Serious or repeated breaches may be regarded as gross misconduct in which case dismissal may be instant and without prior warning.

In addition to its statutory duty to provide an Accident Book, the Company has instituted a system for reporting accidents, diseases, and dangerous occurrences to the Health and Safety Executive under RIDDOR 1995.

In order to meet its obligations to the general public and all lawful visitors to the Company's premises, the Company will pay strict attention to its duties under all health and safety legislation and the Occupiers' Liability Acts of 1957 and 1984.

This Policy has been prepared in compliance with Section 2(3) of the Health and Safety at Work etc Act 1974 and is binding on all Directors, Managers and Employees. All clients, customers and visitors are asked to respect this Policy, a copy of which can be obtained by request

Signed:_____
Date_____
Managing Director

4

The Accident Book

One of the first records you need is an accident book. This is an approved document *required by the HSWA 1974 if you employ ten or more employees.* It is known as a **"BI 510"** and is obtainable from HMSO and some bookshops. A specimen page from an accident book is reproduced on page 33. The entries in the accident book are self-explanatory. They are also confidential under the *Data Protection Act 1998.*

Ideally, the injured person should himself make the entry in the accident book, but if he is unable to do so the entry can be made on his behalf. He should later countersign the entry to show his agreement with the details recorded. You, as his employer, should advise him to apply to his local benefits Agency for a **form B195**, submission of which records that he has sustained an industrial accident. This could be significant in any future claim for benefit. The accident book should also be used for recording accidents involving customers or visitors to your premises. *It should be kept for 3 years.*

Reporting of injuries

Recording details of an accident at work in your accident book is often only the first step in a procedure involving reporting accidents and incidents to the HSE on official forms under the *Reporting of Injuries, Diseases and Dangerous Occurrences Regulations 1995* (RIDDOR).

Specimen page from BI 150 accident book

1 About the person who had the accident
Give full name, home address and occupation.

FULL NAME

ADDRESS

POSTCODE

OCCUPATION

2 About you, the person filling in this book
If you did not have the accident, give full name, home address and occupation.

FULL NAME

ADDRESS

POSTCODE

OCCUPATION

3 Please sign and date *(the person filling in the book)*

SIGNATURE   DATE   /   /

The person who has had the accident should sign and date if they have not filled in the book (as confirmation that they agree the accident recorded is a true and accurate record).

SIGNATURE   DATE   /   /

4 About the accident *When and where it happened.*

DATE   /   /   TIME

IN WHAT ROOM OR PLACE DID THE ACCIDENT HAPPEN?

5 About the accident – what happened
Say how the accident happened. Give the cause if you can. In the event of any personal injury, say what it is.

HOW DID THE ACCIDENT HAPPEN?

MATERIALS USED IN TREATMENT

6 Reporting of injuries, diseases and dangerous occurrences 1995 *(see page iii)*
For the employer only – complete the box provided if the accident is reportable under RIDDOR.

HOW REPORTED

DATE REPORTED   /   /   EMPLOYER'S NAME AND INITIALS

Before tackling RIDDOR, first you need to know the difference between an incident and an accident:

- An **incident**, according to the HSE, *"includes any undesired circumstances and near misses which have the potential to cause accidents"*.

- An **accident** is even more long-windedly described by the HSE as *"any undesired circumstances which give rise to ill-health or injury, damage to property, plant, products or the environment, production losses or increased liabilities"*. RIDDOR itself describes an accident as *"including an act of non-consensual physical violence done to a person at work and an act of suicide occurring in the operation of a transport system."*

Under the Regulations, whenever any of the following **incidents** occur *"out of or in connection with work"*, you must report it to your enforcing authority by the quickest practicable means (e.g. telephone) and in writing within ten days:

- The death of any person as a result of an accident, whether or not he is at work.
- A member of your staff who is at work suffers a major injury as a result of an accident.
- A person who is at not at work (e.g. a member of the public) suffers an injury arising out of or in connection with work and is taken from the scene to a hospital for treatment in respect of that injury.
- A person not at work suffers a major injury as a result of an accident arising out of or in connection with work at a hospital
- One of a list of specified "dangerous occurrences" takes place. (Dangerous occurrences are events which do not necessarily result in a reportable injury, but have the potential to do significant harm – such as a "near miss".)
- A member of your staff is unable to do his normal work for more than three days as a result of an injury (an "over-3-day" injury) caused by an accident at work.

The report must be be made in writing within 10 days by submission to the HSE of a **Form 2508** for accidents and dangerous occurrences and **Form 2508A** for notifiable diseases. It is a simple matter then to photocopy the forms and file them for your own records.

A specimen **Form F2508** is reproduced on page 38 for your information. Both forms can also be downloaded from the HSE website. Once completed, they should be sent to Incident Contact Centre in Caerphilly. (The full address is on the forms.) Important points to be aware of are:

- Accident is now defined to include acts of violence done to people at work and acts of suicide on railways or other relevant transport systems
- Reportable injuries to people not at work (e.g. members of the public) are now deaths, and any injuries which cause a person to be taken from the site of the accident to a hospital.
- The time limit for reporting over three day absences has been increased from 7 to 10 day.

The report required to be sent to the enforcing authority may be sent either on a form approved by the HSE or by some other means approved by the HSE. This enables the HSE to approve the sending of reports by, for instance, telephone.

Records of reportable incidents

You are obliged keep records of reportable incidents for at least three years from the date of entry. Your records must include the date and time of the incident, full details of the person involved, the nature of the injury, disease, etc., the location where the incident occurred, and a full description of the circumstances. You should keep the records on your premises where they can be made available on request to your enforcing authority.

Records should be kept at the place of work and made available on request to enforcing authorities.

Definitions

Major injuries

- Any fracture, other than to the fingers, thumbs or toes.
- Any amputations.
- Dislocation of the shoulder, hip, knee or spine.
- Loss of sight (whether temporary or permanent).
- A chemical or hot metal burn to the eye or any penetrating injury to the eye.
- Any injury resulting from an electric shock or electrical burn (including any electrical burn caused by arcing or arcing products) leading to unconsciousness or requiring resuscitation or admittance to hospital for more than 24 hours.
- Any other injury
  - leading to hypothermia, heat-induced illness or to unconsciousness
  - requiring resuscitation, or
  - requiring admittance to hospital for more than 24 hours.
- Loss of consciousness caused by asphyxia or by exposure to a harmful substance or biological agent.
- Either of the following conditions which result from the absorption of any substance by inhalation, ingestion or through the skin:
  - acute illness requiring medical treatment, or
  - loss of consciousness.
- Acute illness which requires medical treatment where there is reason to believe that this resulted from exposure to a biological agent or its toxins or infected material.

Dangerous occurrences

- Collapse of lifting machinery.
- Failure of pressure systems, e.g. boilers, pipework.
- Failure of freight container.
- Unintentional incident involving overhead electric lines exceeding 200 volts.
- Electrical short-circuit which stops "plant" for more than 24 hours or is potentially life-threatening.

- Incidents involving explosion.
- Incidents involving release or potential release of biological agents likely to cause severe human infection or illness.
- Incidents involving malfunction of radiation generators used in radiography, irradiation of food or processing of products by irradiation.
- Incidents involving breathing apparatus while in use or under test in certain circumstances.
- Incidents in relation to diving operations which put the diver at risk.
- Incidents involving collapse of scaffolding more than 5 metres high or erected over or adjacent to water where there is a risk of drowning.
- Incidents involving train collisions.
- Incidents in relation to wells.
- Incidents in respect of a pipeline or pipeline works.
- Incidents involving failure of fairground equipment in use or under test.
- Incidents involving the carriage of dangerous substances by road.
- Incidents involving unintended collapse of a building or structure involving a fall of more than 5 tonnes of material.
- Incidents in respect of explosion or fire resulting in the stoppage of plant for more than 24 hours.
- Incidents involving the escape of flammable substances.

Diseases

The notifiable diseases include the following:

- Certain poisonings.
- Some skin diseases – occupation dermatitis, skin cancer, chrome ulcer, oil folliculitis and acne.
- Lung diseases, including occupational asthma, farmer's lung, pneumoconiosis, asbestosis and mesothelioma.
- Infections such as; leptospirosis, hepatitis, tuberculosis, anthrax, legionellosis and tetanus.
- Other conditions – occupational cancer, certain musculoskeletal disorders.

Specimen Form F2508

Health and Safety at Work etc Act 1974
The Reporting of Injuries, Diseases and Dangerous Occurrences Regulations 1995

HSE
Health & Safety
Executive

Click here for repo

Report of an injury or dangerous occurrence

Filling in this form
This form must be filled in by an employer or other responsible person.

Part A

About you
What is your full name?

What is your job title?

What is your telephone number?

About your organisation
What is the name of your organisation?

What is its address and postcode?

What type of work does the organisation do?

Part B

About the incident
On what date did the incident happen?

At what time did the incident happen?
(Please use the 24-hour clock eg 0600)

Did the incident happen at the above address?
Yes ☐ Go to question 4
No ☐ Where did the incident happen?
☐ elsewhere in your organisation – give the name, address and postcode
☐ at someone else's premises – give the name, address and postcode
☐ in a public place – give details of where it happened

If you do not know the postcode, what is the name of the local authority?

In which department, or where on the premises, did the incident happen?

Part C

About the injured person
If you are reporting a dangerous occurrence, go to Part F. If more than one person was injured in the same incident, please attach the details asked for in Part C and Part D for each injured person.

1 What is their full name?

2 What is their home address and postcode?

3 What is their home phone number?

4 How old are they?

5 Are they
☐ male?
☐ female?

6 What is their job title?

7 Was the injured person (tick only one box)
☐ one of your employees?
☐ on a training scheme? Give details:

☐ on work experience?
☐ employed by someone else? Give details of the employer:

☐ self-employed and at work?
☐ a member of the public?

Part D

About the injury
1 What was the injury? (eg fracture, laceration)

2 What part of the body was injured?

3 Was the injury (tick the one box that applies)

- [ ] a fatality?
- [ ] a major injury or condition? (see accompanying notes)
- [ ] an injury to an employee or self-employed person which prevented them doing their normal work for more than 3 days?
- [ ] an injury to a member of the public which meant they had to be taken from the scene of the accident to a hospital for treatment?

Did the injured person (tick all the boxes that apply)

- [ ] become unconscious?
- [ ] need resuscitation?
- [ ] remain in hospital for more than 24 hours?
- [ ] none of the above.

Part E

About the kind of accident

Please tick the one box that best describes what happened, then go to Part G.

- [ ] Contact with moving machinery or material being machined
- [ ] Hit by a moving, flying or falling object
- [ ] Hit by a moving vehicle
- [ ] Hit something fixed or stationary

- [ ] Injured while handling, lifting or carrying
- [ ] Slipped, tripped or fell on the same level
- [ ] Fell from a height

  How high was the fall?

  [_____] metres

- [ ] Trapped by something collapsing

- [ ] Drowned or asphyxiated
- [ ] Exposed to, or in contact with, a harmful substance
- [ ] Exposed to fire
- [ ] Exposed to an explosion

- [ ] Contact with electricity or an electrical discharge
- [ ] Injured by an animal
- [ ] Physically assaulted by a person

- [ ] Another kind of accident (describe it in Part G)

Part F

Dangerous occurrences

Enter the number of the dangerous occurrence you are reporting. (The numbers are given in the Regulations and in the notes which accompany this form)

[_____]

Part G

Describing what happened

Give as much detail as you can. For instance

- the name of any substance involved
- the name and type of any machine involved
- the events that led to the incident
- the part played by any people.

If it was a personal injury, give details of what the person was doing. Describe any action that has since been taken to prevent a similar incident. Use a separate piece of paper if you need to.

Part H

Your signature

Signature

[_____]

Date

[_____]

If returning by post/fax, please form is signed, alternatively, if by E-Mail, please type your nam signature box

Where to send the form
Incident Contact Centre, Caerphilly Business Centre, Caerphilly Business Park, Caerphilly, CF83 3GG. or email to riddor@natbrit.com or fax to 0845 300 99 24

For official use

Client number	Location number	Event number	
[_____]	[_____]	[_____]	[ ] INV REP [ ] Y [ ] N

5

Fire Precautions

Fire and emergency procedures

Before 1997, fire safety for business premises was conducted mainly by the local fire authority under the *Fire Precautions Act 1971*. If, at that time, more than 20 persons were employed in your premises at any one time, or more than 10 persons were employed at any one time to work elsewhere than on the ground floor, or your premises were used for the storage of explosives or a highly flammable material, they required a fire certificate issued by the local fire authority. The fire certificate was only issued if the fire authority was satisfied regarding the following: the premises' adequate means of escape; the effective and safe use of the means of escape at all times; the means for fighting fire; and the fire warning system.

The inspection of premises by the fire brigade was both time-consuming and bureaucratic and in 1997 new regulations were introduced to reduce this burden. Under the new regulations, an employer had to carry out a fire risk assessment of his workplace himself, as, according to the Home Office, it should be self-evident that any fire precautions should be based on the level of fire risk for the premises. Those Regulations were further amended in 1999, and in 2005 our untiring legislators excelled themselves yet again by producing more regulations. You, as an employer, must now do your fire risk assessment under the guidance of *The Regulatory Reform (Fire Safety) Order 2005*.

You are already required under the *Management of Health and Safety at Work Regulations 1999* (MHSW) – also known as the **Management Regulations** – to make an assessment of the general risks to health and safety of your employees to which they are exposed whilst at work. That general risk assessment now includes the risks to health and safety arising from fire. The fire risk assessment may be made as part of the general risk assessment, or it may be done as a separate exercise. Because of the more specialised nature of the subject, you would be well advised to carry out a separate fire risk assessment.

You can obtain a Fire Risk Assessment Record Book from the Fire Protection Agency.

Your fire risk assessment should be based on the following five steps:

1. Identify potential fire hazards
- Sources of ignition
- Sources of fuel
- Sources of oxygen

2. Identify people at risk
- People in and around the premises
- People especially at risk

3. Evaluate, remove, reduce and protect from risk;
- Evaluate the risk of a fire occurring
- Evaluate the risk to people from fire
- Remove or reduce the hazards
- Remove or reduce the risks to people

with particular regard to:

- Detection and warning
- Fire-fighting
- Escape routes
- Lighting
- Signs and notices
- Maintenance

4. Record, plan, inform, instruct and train
- Record significant finding and action taken
- Prepare an emergency plan

- Inform and instruct relevant people; co-operate and co-ordinate with others
- Provide training

5. Review
- Keep assessment under review
- Revise where necessary

(An action checklist covering the first three steps is set out on page 45.)

Under the Order, your workplace must be equipped with *appropriate* fire-fighting equipment and with fire-detectors and alarms. Any non-automatic fire-fighting equipment you provide must be easily accessible, simple to use and indicated by signs.

What is "appropriate" equipment?

This is determined by the dimensions and use of the workplace, the equipment it contains, the physical and chemical properties of the substances likely to be present, and the maximum number of people that may be present at any one time.

Fire-fighting measures

Fire-fighting measures must be adapted to the size of your workplace and the nature of your activities.

Don't forget to take into account the presence of persons other than your employees.

If you employ a number of people select and train an adequate number of them to implement your fire-fighting measures. Make sure that their training and the available equipment e.g. fire extinguishers, hose reels, sprinklers systems, etc. are adequate based on the size of your workplace and the specific hazards contained in it.

In addition, you must arrange contacts with the external emergency services, particularly as regards rescue work and fire-fighting.

Classification of fires

Fires are classified according to what is actually burning as shown in the following table:

Class of fire	Description
Class A	Fires involving solid materials such as wood, paper or textiles
Class B	Fires involving flammable liquids
Class C	Fires involving gases
Class D	Fires involving metals
Class F	Fires involving cooking oils such as in deep-fat fryers

(If there is a possibility of a Class C or D fire on your premises get advice from a competent person)

Fire extinguishers

All portable fire extinguishers are red with their contents indicated by a coloured band round the body. You can then buy appropriate extinguishers for the potential class of fire likely to occur in your premises. An illustration of the different types of extinguisher available is set out on page 44.

You must have a suitable system of maintenance for any equipment and devices you provide for your workplace and they must be kept in an efficient state and in a good state of repair.

Types of fire extinguisher

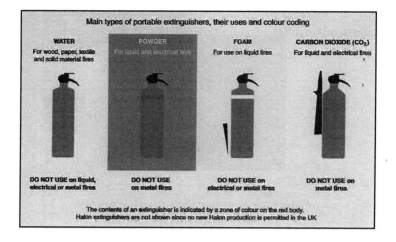

Main types of portable extinguishers, their uses and colour coding

WATER	POWDER	FOAM	CARBON DIOXIDE (CO₂)
For wood, paper, textile and solid material fires	For liquid and electrical fires	For use on liquid fires	For liquid and electrical fires
DO NOT USE on liquid, electrical or metal fires	DO NOT USE on metal fires	DO NOT USE on electrical or metal fires	DO NOT USE on metal fires

The contents of an extinguisher is indicated by a zone of colour on the red body.
Halon extinguishers are not shown since no new Halon production is permitted in the UK

In case of fire

In case of fire the following factors are crucial:

- Routes to emergency exits from your workplace and the exits themselves are kept clear at all times.
- Emergency routes and exits must lead as directly as possible to a place of safety.
- Employees must be able to evacuate the workplace quickly and as safely as possible.
- The number, distribution and dimensions of emergency routes and exits must be adequate having regard to the use, equipment and dimensions of the workplace and the maximum number of persons present there at any one time.
- Emergency doors must open in the direction of escape.
- Sliding or revolving doors must not be used for exits specifically intended as emergency exits.
- Emergency doors must not be so locked or fastened that they cannot be easily and immediately opened by any person who may require to use them in an emergency.
- Emergency routes and exits must be indicated by signs.

- Emergency routes and exits which are normally illuminated must be provided with adequate emergency lighting in the case of failure of their normal lighting.

If your failure, either intentionally or recklessly, to comply with any of the provisions of the Regulations places one or more of your employees at serious risk of death or serious injury from fire you are liable in law.

A specimen fire risk checklist and fire risk assessment are reproduced below as examples but you can produce your own forms provided they contain relevant and significant findings and recommendations.

Fire risk assessment checklist

1. Identifying the fire hazards

Sources of Ignition

1.	Does the work involve hot processes such as incineration, welding, flame-cutting, cooking or the use of ovens?	Yes/No
2.	Are there light-bulbs and fittings near combustible materials?	Yes/No
3.	Are there fluorescent light-bulbs and fittings near combustible materials?	Yes/No
4.	Are portable heaters (gas, electric, oil-fired) in use?	Yes/No
5.	Are there multi-point adaptors in electrical sockets?	Yes/No
6.	Are electrical extension leads plugged into adaptors or other extension leads?	Yes/No
7.	Are any portable electrical appliances faulty or damaged?	Yes/No

8. Are there any faults with the electrical installation?	Yes/No
9. Is cooking carried on in the premises?	Yes/No
10. Is arson a potential problem?	Yes/No
11. Are there any other potential sources of heat in the workplace?	Yes/No
If Yes, what/where are these sources of heat?	

Sources of Fuel

1. Do the work processes involve the use of combustible materials?	Yes/No
2. Are bulk quantities of combustible materials present?	Yes/No
3. Are large amounts of textiles and furniture present?	Yes/No
4. Are there large amounts of foam padding displayed or stored in the workplace?	Yes/No
5. Are items of furniture damaged with padding exposed?	Yes/No
6. Are there large notice boards on escape routes?	Yes/No
7. Is more than 20% of the walls covered with combustible linings such as hardboard, chipboard, plastic tiles or flock wallpaper?	Yes/No
8. Is there any area of the walls covered with carpet tiles?	Yes/No
9. Is the ceiling covered with polystyrene tiles?	Yes/No

10. Are combustible artificial foliage or plants present?	Yes/No
11. Are there displays involving combustible materials and panels?	Yes/No
12. Are paper or similar decorations hung in the building during Christmas or other festival times?	Yes/No
13. Are there any other combustible materials in the workplace?	Yes/No
If Yes, what and where are they?	
14. Are stocks of flammable liquids for use in the processes stored in the workplace?	Yes/No
15. Are containers of flammable liquids left open without their tops on?	Yes/No
16. Are flammable liquids kept in the workplace for use by the cleaners or maintenance staff?	Yes/No
17. Are quantities of flammable liquids kept for any other purposes?	Yes/No
If Yes, what are they and where are they kept?	
18. Is natural gas used in any manufacturing process?	Yes/No
19. Are cylinders of flammable gases used or stored on the premises?	Yes/No
20. Are cylinders of other gases, such as air or oxygen, used or stored in the premises?	Yes/No
21. Are aerosol cans filled or stored in the premises?	Yes/No
22. Are there any other forms of compressed gas used in storage?	Yes/No

If yes, what are they and where are they stored?	

Sources of Oxygen

1.	Are stocks of raw materials and finished products separated from the workplace by a fire-resistant structure?	Yes/No
2.	Are the compartments enclosed by a fire-resistant structure?	Yes/No
3.	Are all holes in compartment walls, ceilings and floors around services such as pipes and cables fire-stopped?	Yes/No
4.	Have dampers been installed in ductwork where it passes through compartment walls, ceilings and floors?	Yes/No
5.	Are holes in the floors and ceilings of vertical service ducts or cupboards fire-stopped?	Yes/No
6.	Are all openings in compartment boundaries protected in case of fire?	Yes/No
7.	Are there undivided floor voids?	Yes/No
8.	Are there undivided ceiling voids?	Yes/No
9.	Are there voids behind panelling or other features which could lead to fire-spread in the floor above?	Yes/No
10.	Are there any other features which could lead to the spread of flames or smoke in the event of fire?	Yes/No
11.	Are any oxidising materials or cylinders stored in the building?	Yes/No
	If Yes, give location of voids and location of any other hazardous features	

2. *Identifying people at risk*

1.	Is there a large number of staff in the workplace?	Yes/No
2.	Are immediate neighbours to the premises at risk?	Yes/No
3.	Do members of the public visit the workplace?	Yes/No
4.	Will people be unfamiliar with the layout of the workplace?	Yes/No
5.	Are disabled people on the premises?	Yes/No
6.	Are elderly people on the premises	Yes/No
7.	Are children on the premises?	Yes/No
8.	Do people work in remote areas of the premises?	Yes/No
9.	Are contractors and maintenance workers unaware of the dangers posed by fire?	Yes/No
10.	Do any staff work in areas where there is a high risk of a fire occurring?	Yes/No
11.	Do people sleep in the workplace?	Yes/No
12.	Are there any other factors which put people in the building at risk?	Yes/No
	If Yes, what are they?	

3. *Evaluate, remove reduce and protect from risk*

1.	Can the work process be replaced with one that reduces the potential for ignition? (e.g. replace a hot work method with one that does not require flames or heat)	Yes/No
2.	Can a hot work permit system be introduced?	Yes/No
3.	Can light units be re-positioned to reduce the risk of contact with Combustible materials?	Yes/No
4.	Can radiant heaters and those employing flames with fixed convector heaters or central heating be replaced?	Yes/No
5.	Can additional electrical sockets be installed?	Yes/No
6.	Can protective devices such as RCD's and thermostats be installed and maintained?	Yes/No
7.	Can electrical wiring and portable appliances be inspected regularly?	Yes/No
8.	Can appropriate security measures against arson be introduced?	Yes/No
9.	Can any combustibles be replaced with non-combustible alternatives?	Yes/No
10.	Can the amount of combustible materials stored in the premises be reduced?	Yes/No
11.	Can combustible materials be stored away from potential sources of ignition?	Yes/No
12.	Can the amount of materials that are being displayed be reduced?	Yes/No
13.	Can furniture made from combustible upholstery be replaced with furniture which is less combustible?	Yes/No
14.	Can damaged furniture be replaced?	Yes/No
15.	Can housekeeping and the arrangements for the disposal of waste and rubbish be improved?	Yes/No
16.	Can combustible wall linings be replaced with more suitable materials?	Yes/No

17. Can combustible ceiling linings be replaced or removed?	Yes/No
18. Can the size of notice boards and the amount of paper notices be reduced?	Yes/No
19. Can the hanging of combustible decorations near light fittings be prohibited?	Yes/No
20. Can the volume of flammable liquids kept in the workplace be reduced?	Yes/No
21. Can all containers be kept closed when not in use?	Yes/No
22. Can flammable liquids or solvents be replaced with non-flammable alternatives?	Yes/No
23. Can flammable liquids used by cleaners and maintenance staff be removed/reduced/replaced?	Yes/No
24. Is gas-filled equipment serviced and maintained regularly?	Yes/No
25. Can the number of cylinders of flammable and non-flammable gases kept in the workplace be reduced?	Yes/No
26. Can the number of aerosol cans stored in the premises be reduced/replaced with less hazardous products?	Yes/No
27. Where storage of large quantities of aerosols is unavoidable, are they stored in purpose-built cages?	Yes/No
28. Are stocks of raw materials and finished products separated from the Workplace by means of a fire-resistant structure?	Yes/No
29. Are voids beneath floors separated?	Yes/No
30. Are ceiling voids divided?	Yes/No
31. Are all holes around services fire-stopped to the same standard as the fire resistance of the element of construction in which they are situated?	Yes/No
32. Can automatically-operating fire-resistant doors or shutters be installed to protect openings in compartment walls?	Yes/No

33. If people sleep in the workplace: (a) Is there an early warning of fire? (b) Have sleeping areas been evacuated?	Yes/No Yes/No
34. If large numbers of people are present, especially members of the public, is there a sufficient number of trained staff to ensure speedy and orderly evacuation?	Yes/No
35. If the workplace is regularly used by persons with impaired mobility: (a) Is the number of trained staff adequate to ensure safe evacuation? (b) Are the escape routes are suitable for the persons who have to use them?	Yes/No Yes/No
36. If the layout and escape routes may be unfamiliar to the people present: (a) Are the escape routes adequately signed? (b) Is the number of trained staff adequate to ensure safe evacuation? (c) Can instructions and advice be given by a voice alarm or public address system?	Yes/No Yes/No Yes/No
37. If people present may be unaware of dangers posed by fire, have adequate arrangements been made for their safe evacuation?	Yes/No
38. If people at work are exposed to a high risk of fire: (a) Have they been trained appropriately for the hazards? (b) Have they been trained in the action to take in the event of fire?	Yes/No Yes/No
39. In the case of small workplaces, can the work activity be arranged so that any outbreak of fire can be seen immediately by people present?	Yes/No
40. Can an automatic fire detection and alarm system be provided?	Yes/No
41. Can an automatic sprinkler or other suitable fixed fire-fighting system be provided?	Yes/No
42. Can a smoke control system be provided?	Yes/No
43. Can the source of ignition be contained by providing fire-resisting walls, doors or shutters?	Yes/No

Fire risk assessment

By extracting the relevant positive and negative responses from the above checklist, you will be able to form the basis of the risk assessment itself. An example of an appropriate, simple risk assessment is set out on page 54.

Specimen Fire Risk Assessment

J Smith Ltd 100 High Street London	Date: 01/01/2008 Assessor J Smith Signature_____
Ground Floor:	One large open-plan office, small kitchen, staff toilet Business: accountancy
xxxxxxxxxxxxxxxxxxxxxxxxxxxxxxxxxxxxxxxxxxxxx	
Fire Hazards	
Sources of ignition	Lighting, 2 computers, 1 printer, 1 microwave oven, 1 small electric cooker.
Sources of fuel	Wooden Office furniture, carpet tiles, paper, files, stationery.
Sources of oxygen	N/A apart from natural ventilation.
People at risk	4 full-time members of staff.
Xxxxxxxxxxxxxxxxxxxxxxxxxxxxxxxxxxxxxxxxxxxxx	
Evaluate, remove, reduce and protect from risk	
Evaluate risk of fire occurring	No smoking policy; PAT testing of all electrical appliances, good housekeeping procedure in operation: low risk of fire.
Evaluate risk to people from a fire starting on	All staff are able-bodied; door into kitchen has premises large vision panel.
Remove and reduce hazards	N/A: hazards present are necessary for conduct.
Remove and reduce the risk	Obtain additional fire extinguishers. Keep rear fire-exit door free from obstruction.
Review Date	
Completed by_____Signature_____Date_____	

The Regulatory Reform (Fire Safety) Order 2005

6

First Aid

The *Health and Safety (First-Aid) Regulations 1981* require employers to provide adequate and appropriate equipment, facilities and personnel to enable first aid to be given to employees if they are injured or become ill at work. These Regulations apply to all workplaces including those with five or fewer employees and to the self-employed.

The *minimum* first-aid provision on any work site is:

- a suitably stocked first-aid box (see Q4);
- an appointed person to take charge of first-aid arrangements.

An assessment has to made as to how much you need to provide in the way of first aid facilities based on the number of employees, work activities, access to local hospitals, employees working away from base, and first aid cover for absences due to sickness or holidays. Your assessment will determine how many trained first aiders are required, the number of first aid boxes needed, and whether you require a first aid room, stretchers or carrying chairs. You can select a few members of staff to obtain first aid training and qualifications at your expense or have a number of "appointed persons" depending on the size and nature of your business. **Qualified first aiders** must have attended a recognised course and have become accredited first aiders for a period of 3 years. **Appointed persons** are members of staff selected to look after the first aid box and summon the ambulance if required. For first aid purposes premises are categorised as low, medium and high risk. The following table gives the current HSE guidelines for the numbers of first aiders required:

HSE: Suggested numbers of first aid personnel

Number of staff	Number of first aiders required
1. Low Risk *(Offices, shops, etc)*	
Less than 50	1 appointed person
50-100	1 first aider
100+	1 extra first aider for every 100 staff
2. Medium Risk *(Light engineering, hotels, warehouses, recreational premises)*	
Less than 20	At least 1 appointed person
20-50	At least 1 first aider
51-100	At least 2 first aiders
More than 100	One extra first aider for every 100 staff
3. High Risk *(Agriculture, construction, chemical plants)*	
Less than 5	At least 1 appointed person
5-50	At least 1 first aider
Over 50	1 extra first aider for every 50 staff

For further advice on first aid in the workplace, see *First Aid in the Workplace* by Eva Roman (also published Management Books 2000).

7

Risk Assessment

The old adage says that *prevention is better than cure* and our legislators have set out the following general principles of prevention:

- Avoid risks.
- Evaluate those risks which cannot be avoided.
- Combat the risks at source.
- Adapt the work to the individual.
- Adapt in the light of new technology.
- Replace the dangerous by the less dangerous (preferably non-dangerous).
- Develop an overall prevention policy embodying technology, work organisation, working conditions, the working environment, and social relationships.
- Prioritise collective protective measures over individual protection.
- Give appropriates instructions and training to employees

Risk avoidance is the gold standard but life and work is full of risks so you are faced with evaluation or "risk assessment". Several rainforests have been destroyed to provide the dozens of books and articles written on this subject. Safety consultants have made a handsome living out of it by making risk assessment sound like some mysterious process, understood by only a chosen few experts, in order to bamboozle an unsuspecting fee-paying public.

Risk is inherent in living. All forms of travel by train, boat, plane or car involve risk. Playing most sports involve risk. We look at the risks in these activities and weigh them (consciously or unconsciously) against the costs, the benefits derived from the activity, and the likelihood of the risks putting us in danger. So it is not a mysterious process – it is a simple process which we all perform on numerous occasions as a matter of course, e.g. driving a car or crossing the road. In health and safety terms, the main difference is that under certain circumstances you are required to write down your risk assessment and make it available to your employees.

Take a closer look at what is involved. You, as an employer, must assess the risk to the health and safety of your employees and anyone else affected by your work activity. Following on from your risk assessment you must implement the preventive and protective measures your assessment identified as being necessary. *Your risk assessments should be written and kept under review.*

So what does a risk assessment consist of and where do we start? The first step is to look at your business activities and decide if any of them are in themselves hazardous. Before you start you need to understand the terms hazard and risk.

What is a hazard?

A hazard is anything which has the potential to cause harm. When crossing the road, the approaching car is a hazard because it can harm you. So what is hazardous in your business? If your business is office-based, we can list the activities in your office to see if there is anything there that can cause harm: word-processing, filing, e-mailing, telephoning... Nothing terribly hazardous there, then. Or is there? Take word-processing. It involves the use of a computer, keyboard and mouse. Staring at the computer screen might give you sore eyes or a headache. Prolonged use of the keyboard and mouse can give you pains in your hands and arms.

Having identified the hazards, the next step is to evaluate the risk to health and safety posed by those hazards.

What is risk?

Risk is the chance or likelihood that harm from a particular hazard might be realised. In the road-crossing analogy there is a risk of being knocked down if you have misjudged the speed of the approaching car or if the driver swerves and mounts the pavement. Getting back to your office environment, it is well established that incorrect use of a keyboard and/or mouse may cause pain to the hand, arm, neck and back, known as work-related upper limb disorder – so there is a risk of sustaining injury from the use of the keyboard or incorrect posture

The purpose of risk assessment is to identify the steps which must be taken in order to control any health and safety risks to employees and others arising from your business. The assessment should be carried out by a **"competent person"**, and enable you to set out your health and safety priorities.

What is a "competent person"?

The Regulations define a competent person as someone with sufficient training, experience, knowledge or other qualities to assist him in complying with the requirements and prohibitions imposed by the relevant legislation. You don't need particular skills or qualifications to be competent in health and safety terms. Put simply, most situations require only an understanding of relevant current best practice; an awareness of the limitation of your own experience and knowledge; and the will and ability to acquire the necessary experience and knowledge. Once you have appointed your competent persons – health and safety representatives, fire marshals, first aiders, etc., you need to tell your employees who and where they are.

We mentioned hazards associated with offices and display screen equipment (DSE) but hazards in workplaces come in all shapes and sizes. Examples include:

- lifting heavy or awkward loads
- using tools and machinery

- working at height
- working in confined spaces
- being exposed to loud or sudden noise
- working in extremes of temperature
- working on badly managed sites
- working with chemicals.

You can identify most hazards by simply walking around your workplace and making careful observations ("Management by walking about" – a good management maxim). It is also sensible to involve your staff and ask their opinions. They probably know the problems better than you do! Once the hazards have been identified you need to list the risks arising from them. You then need to evaluate them and put them into some order of priority. A simple way is to rate each risk as **high**, **medium** or **low**. This in turn allows you to prioritise your remedial measures.

Back to risk assessment! How do you get to grips with your risk assessment problem? The Regulations don't actually explain how to do a risk assessment so there are no fixed rules. The assessment can range from the very simple, based on straightforward ordinary common sense, to the highly technical, based on a professional qualified risk assessment. It may be generic assessment or based on a variety of separate assessment exercises. "Generic" simply means that, for example, if you have ten identical workstations with identical equipment and accessories, you can base your assessment of the workstations on one of the ten and it will apply across the board. Where you have carried out an assessment under one set of regulations, you don't have to repeat it under other regulations so long as it remains valid. For example, both the Workplace Regulations and the DSE Regulations require you to assess your work-stations. One set of assessments will suffice for each!

One very simple method is to assign a numerical weighting to the categories of **severity** and **likelihood** as shown below.

Likelihood		Severity	
Certainty	5	Fatality	5
Very likely	4	Major injury, disablement, disease	4
Likely	3	Minor injury	3
Unlikely	2	Damage to plant	2
Very unlikely	1	Delay/interruption	1

We then multiply the factors to ascribe combined risk ratings as displayed in the following matrix. (For example, the rating for a Certain Fatality is 5 x 5 = 25.) The higher the risk rating the higher, the greater the priority.

Severity/Likelihood Matrix

	Certain (5)	Very likely (4)	Likely (3)	Unlikely (2)	Very unlikely (1)
Fatality (5)	25	20	15	10	5
Major injury, disease, disablement (4)	20	16	12	8	4
Minor injury (3)	15	12	9	6	3
Damage to plant (2)	10	8	6	4	2
Delay/ interruption (1)	5	4	3	2	1

Based on this method of risk prioritisation, we can determine the appropriate "reasonably practicable" actions. The assessment of risk should be weighed against the time, trouble, cost and difficulty of doing anything about it. The following descending scale awards priority in relation to the numerical value of the risk.

Risk Factor Prioritisation

Risk factor	Prioritisation
25	Very serious risk – *take immediate action*
15-25	Risk not adequately controlled – re-evaluate risk assessment and implement appropriate controls
10-14	Compare with current standards
5-9	Adequately controlled risk – no further action required
1-4	Trivial acceptable risk – no action required

Obviously, a fatality even in the "very unlikely" category would require immediate action to be taken. At the other end of the spectrum, if the possibility of a minor injury is unlikely, then that may be classed as an "acceptable" risk. Remember, however, that a succession of relatively minor injuries could also push the problem up the priority scale. You need to decide whether your existing safeguards or control measures, if any, are adequate or whether you need to introduce more safety measures. You might eliminate the hazard completely, substitute the activity with a less hazardous one, or reduce the number of people exposed to the hazard.

Recording the assessment

Your next step is to record the assessment. *If you have five or more employees you are legally obliged to record the assessment.* Ideally, it should be linked to your health and safety policy.

You can record it electronically so long as it is readily retrievable.

You need to include at least the following information:

- a description of the process/activity assessed
- identification of the significant risks
- identification of any group of workers particularly at risk

- date of assessment and, where appropriate, the date of the next review
- the name of the competent person carrying out the assessment.

Next, you need to give the relevant details of the assessment and the risks you have identified to all workers who may be affected.

Regulations governing risk assessments

HSE guidance, suppliers' instructions, and trade press all contain information relevant to your particular organisation and you will find them a very useful resource when dealing with risk assessments. Appendix 2 contains details of HSE publications.

Risk assessment is included in a number of health and safety regulations and in some cases there is some overlapping. Where this occurs, you don't need to do a risk assessment for both. The following Regulations require risk assessments to be carried out:

- *The Management of Health and Safety at Work Regulations 1992/1999*
- *The Provision &Use of Work Equipment Regulations 1992/1998*
- *The Workplace (Health, Safety & Welfare) Regulations 1992*
- *The Manual Handling Operations Regulations 1992*
- *The Personal Protective Equipment at Work Regulations 1992*
- *The Health and Safety (Display Screen Equipment) Regulations 1992*
  (The above are sometimes referred to as the "Six-Pack"as they all appeared in 1992.)
- *The Control of Substances Hazardous to Health Regulations 2002*
- *The Regulatory Reform (Fire Safety) Order 2005*

Other regulations concerning asbestos, lead, noise and construction work require risk assessment but are unlikely to feature prominently in small business operations.

A simple but effective method for carrying out risk assessment is to use the HSE's "Five Steps to Risk Assessment" reproduced here:

Five steps to risk assessment

Company name: _____   **Date of risk assessment:** _____

Step 1
What are the hazards?

Spot hazards by:
- walking around your workplace;
- asking your employees what they think;
- visiting the your industry areas of the HSE website or calling the HSE Infoline;
- calling the Workplace Health Connect Adviceline or visiting their website;
- checking manufacturers' instructions;
- contacting your trade association.

Don't forget long-term health hazards.

Step 2
Who might be harmed and how?

Identify groups of people. Remember:
- some workers have particular needs;
- people who may not be in the workplace all the time;
- members of the public;
- if you share your workplace think about how your work affects others present.

Say how the hazard could cause harm.

Step 3
What are you already doing?

List what is already in place to reduce the likelihood of harm or make any harm less serious.

What further action is necessary?

You need to make sure that you have reduced risks 'so far as is reasonably practicable'. An easy way of doing this is to compare what you are already doing with good practice. If there is a difference, list what needs to be done.

Step 4
How will you put the assessment into action?

Remember to prioritise. Deal with those hazards that are high-risk and have serious consequences first.

Action by whom   Action by when   Done

Step 5 Review date:
- Review your assessment to make sure you are still improving, or at least not sliding back.
- If there is a significant change in your workplace, remember to check your risk assessment and, where necessary, amend it.

Five steps to risk assessment

Step 1. Identify the hazards

Look only for hazards which you could reasonably expect to result in significant harm under the conditions in your workplace. Use the following examples as a guide:

- Slipping/tripping hazards (e.g. poorly maintained floors or stairs)
- Fire (e.g. from flammable materials)
- Chemicals (e.g. battery acid)
- Moving parts of machinery (e.g. blades)
- Work at height (e.g. from mezzanine floors)
- Ejection of materials (e.g. from plastic moulding)
- Pressure systems (e.g. steam boilers)
- Vehicles (e.g. fork-lift trucks)
- Electricity (e.g. poor wiring)
- Dust (e.g. from grinding)
- Fumes (e.g. welding)
- Manual handling
- Noise
- Poor lighting
- Low temperature

Step 2. Decide who might be harmed and how

There is no need to list individuals by name – just think about groups of people doing similar work or who may be similarly affected by the same risks, e.g.

- Office staff
- Maintenance Personnel
- Contractors
- People sharing your workplace
- Operators
- Cleaners
- Members of the public

Pay particular attention to:

- Staff with disabilities
- Visitors
- Inexperienced staff
- Lone workers

They may be more vulnerable.

Step 3. Evaluate the risks and decide on precautions

Is more needed to control the risk?
  For the hazards listed, do the precautions already taken:

- Meet the standards set by a legal requirement?
- Comply with a recognised industry standard?
- Represent good practice?
- Reduce risk as far as reasonably practicable?

Step 4. Record your findings and implement them

Have you provided:

- Adequate information, instruction or training?
- Adequate systems or procedures?

If so, then the risks are adequately controlled, but you need to indicate the precautions you have in place. (You may refer to procedures, company rules, etc.)
  Where the risk is not adequately controlled, indicate what more you need to do.

Step 5. Review your assessment and update if necessary

Set a date for review of the assessment.
  On review check that the precautions for each hazard still adequately control the risk.
  If not, indicate the action needed. Note the outcome. If necessary, complete a new page for your risk assessment.

Make changes in your workplace, e.g. when bringing in:

- new machines
- new substances
- new procedures

These may introduce significant new hazards.
Look for them and follow the five steps for the new hazards.

Pregnant employees

As an employer you need to be particularly aware of risk assessment provisions for women employees of childbearing age or new or expectant mothers. They may be at risk because of a specific work process, working conditions, or exposure to a physical, chemical or a biological agent. Any such risk assessment must identify the specific risk posed to the health and safety of pregnant women and new mothers in the workplace. According to advice from the HSE, the main risks are:

- **Physical agents** – shocks, vibration, manual handling, noise, non-ionising radiation, and extremes of heat and cold.
- **Biological agents** – listeria (bacteria from cheese, milk, coleslaw etc), rubella (German Measles) and chickenpox virus, toxoplasma (infection from eating undercooked meats from infected animals), cytomegalovirus (Herpes virus resembling glandular fever), hepatitis B, and HIV.
- **Chemical agents** – mercury, antimiotic drugs, carbon monoxide, and chemicals listed under certain Directives.
- **Working conditions** – conditions such as mining work and display screen equipment. Although it is widely accepted that work with VDUs does not give rise to problems associated with radiation, your pregnant worker may not accept this. Insistence on your part that she has to continue to work with the equipment could give rise to stress – yet another minefield for you as an employer! You should therefore treat this situation with some sympathy.

Having identified any risks and told your employees about them, you must then remove the hazard or reduce the risk to its minimum. If there is a residual risk, you must temporarily adjust the pregnant employee's working conditions or hours or offer her suitable alternative work. If neither of these alternatives is acceptable, you should suspend the employee on full pay for as long as is necessary to protect her health and safety or that of her child. You should be aware that the "maternity" provisions of the Regulations apply to employees regardless of length of service with an employer.

Young persons

A **young person** is a person who has not attained the age of 18 years (whilst a **child** is one who is not over compulsory school age). If you employ anyone between 15 and 18 years of age, even on a part-time or temporary basis, you need to do a work-placement risk assessment in respect of each one. The following form or something similar is adequate for the purpose.

WORK PLACEMENT RISK ASSESSMENT

Company Name: ....

Address: ....

Trainee's Name: ....

Address: ....

Date of Birth: ....
(If under 16 years, copy of assessment to be provided to parent or guardian)

Trainee's Job Title: ....

Nature of Work: ....

	Yes/No
1. Will the young person be fully covered by the employer's general risk assessment under Regulation 3 of the Management of Health and Safety at Work Regulations 1992?	

---

2. Will the young person be particularly at risk because of any of the following factors?	
Inexperience	Yes/No
Lack of awareness of health and safety requirements	Yes/No
Layout of workplace/workstation	Yes/No
Use and handling of work equipment	Yes/No
Organisation of processes and activities	Yes/No
Exposure to biological, chemical or physical agents	Yes/No
Exposure to ionising radiation	Yes/No
Exposure to high pressure atmospheres	Yes/No
Exposure to sensitising substances by contact or inhalation	Yes/No
Exposure to lead or lead compounds	Yes/No
Exposure to asbestos	Yes/No
Work involving manufacturing/handling of explosive/fireworks	Yes/No
Work involving fierce or poisonous animals	Yes/No
Work involving industrial animal slaughtering	Yes/No
Work involving handling of equipment for production/storage and use of compressed, liquefied or dissolved gases	Yes/No
Work in vats, tanks, reservoirs or carboys containing any of the above substances	Yes/No
Work involving risk of collapsing structures	Yes/No
Work involving high voltage electricity	Yes/No
Work where pace of work is governed by machinery and payment is based on results	Yes/No
Work beyond his/her physical capacity*	Yes/No
Work beyond his/her psychological capacity*	Yes/No
Exposure to toxic substances*	Yes/No
Exposure to carcinogenic substances*	Yes/No
Exposure to substances causing harm to children*	Yes/No
Exposure to substances causing a chronic health effect*	Yes/No
Exposure to extreme temperatures*	Yes/No
Exposure to noise*	Yes/No
Exposure to vibration*	Yes/No
Young persons must not be employed under these conditions unless they are under training, where the work is carried out under the supervision of competent persons, and where the risk has been reduced to the lowest possible level.	

3. Where the answers to any of the above are in the affirmative give details of the control measures in force:

    Existing control measures

    ....

    ....

    ....

    ....

    Proposed control measures

    ....

    ....

    ....

    ....

I certify that the work situation so far as is reasonably practicable does not expose the young person named above to any reasonably foreseeable risk.

Name: ....

Position: ....

Signature: ....

Date: ....

Dangerous substances

It can't be emphasised often enough that there is no need to get bogged down in technicalities or mathematical formulae or employ expensive consultants when dealing with basic risk assessments. They can be dealt with quite simply by the application of common sense and experience. However, you will need professional assistance in carrying out assessments relating to highly dangerous substances such as asbestos and lead.

8

Personal Protective Equipment (PPE)

The rules governing the provision of protective equipment are set out in the *Personal Protective Equipment Regulations 1992.*

Within the terms of these rules "personal protective equipment", or PPE, is defined as "all equipment (including clothing affording protection against the weather) which is intended to be worn or held by a person at work and which protects him against one or more risks to his health or safety, and any addition or accessory designed to meet that objective."

PPE includes aprons, clothing for adverse weather conditions, gloves, safety footwear, safety helmets, high visibility waistcoats, eye protectors, life jackets, respirators, underwater breathing apparatus and safety harnesses.

As an employer you must provide personal protective equipment *at your expense* to any of your employees who may be exposed to health or safety risks while at work. The only exception is where you have controlled the risk suitably by other means.

Before choosing any PPE, you need to assess its suitability for the purpose. This should take into account the following:

- ergonomic requirements and the employee's state of health
- correct fit, if necessary after adjustments
- So far as is reasonable practicable, effectiveness in preventing or controlling the risk or risks involved
- compliance with any relevant legislation relating to design or manufacture with respect to health or safety.

Having issued PPE you must train your staff in its use, ensure that it is properly maintained, and provide suitable storage for it when it is not in use

Types of equipment

Head protection

- Crash helmets, cycling helmets and climbing helmets which are intended to protect the user in falls.
- Industrial scalp protectors (bump caps) which can protect against striking fixed obstacles, scalping or entanglement, and
- Caps, hairnets, etc., which can protect against scalping/ entanglement.

Eye protection

- Spectacles, eye shields, goggles, welding filters, face-shields and hoods. Safety spectacles can be fitted with prescription lenses if required. Some types of eye protection can be worn over ordinary spectacles if necessary.

Foot protection

- Safety boots or shoes.
- Clogs.
- Wellington boots.
- Anti-static footwear prevents the build-up of static electricity on the wearer. It reduces the danger of igniting a flammable atmosphere and gives some protection against electric shock.
- Conductive footwear also prevents the build-up of static electricity. It is particularly suitable for handling sensitive components or substances (e.g. explosive detonators). Does not protect against electric shock.

Hand and arm protection

This include gloves of various designs for protection against:

- Cuts and abrasions.
- Extremes of temperature, hot and cold.
- Skin irritation and dermatitis.
- Contact with toxic or corrosive liquids.

The type and degree of protection depends on the glove material and the way in which it is constructed. Barrier creams may sometimes be used as an aid to skin hygiene in situations where gloves cannot be used. However, barrier creams are less reliable than suitable gloves as a means of chemical protection.

Body protection

- Coveralls, overalls and aprons to protect against chemicals and other hazardous substances.
- Outfits to protect against cold, heat and bad weather.
- Clothing to protect against machinery such as chain-saws.

Protection of the person

- High visibility clothing.
- Life-jackets and buoyancy aids.

Last resort principle

In protecting employees in the workplace, there is in effect an accepted hierarchy of control measures. Engineering controls and safe systems of work should always be considered first, and PPE should always be regarded as the "last resort". It may be possible to do the job by another method which will not require the use of PPE, or, if that is not possible, to adopt other more effective safeguards.

There are a number of reasons for this approach. First, PPE protects only the person wearing it whereas measures controlling the risk at source protect everyone in the workplace. Secondly, theoretical maximum levels of protection are seldom achieved with PPE in practice, and the actual level of protection is difficult to assess. Effective protection is only achieved by suitable PPE, correctly fitted, regularly maintained and properly used. Thirdly,

PPE may restrict the wearer to some extent by limiting mobility or visibility, or by requiring additional weight to be carried. Other means of protection should therefore be used whenever reasonably practicable.

A good example of this is noise in the workplace. Enclosing a noisy machine to reduce or muffle the sound is more effective – and cheaper – than giving individual ear protectors to staff working in the vicinity.

9

Working Conditions

The rules governing the provision of adequate working conditions in the workplace are set out in the *Workplace (Health, Safety and Welfare) Regulations 1992.*

Definition of the workplace

Health and safety at work focuses on what happens in the workplace, so what is a "workplace"? The answer is that it is any premises you make available to your employees as a place of work. It includes any room, lobby, corridor, staircase, road or other place used as a means of access to or egress from the workplace. It also means any place where an employee works such as a yard, scaffolding or even a fork-lift truck.

Clearly some places are self-evidently workplaces – workshops, laboratories, restaurant kitchens, etc. – but so too is your office or shop. Factories, shops, offices, schools, hospitals, nursing homes, hotels, places of entertainment, etc., are all workplaces covered by the workplace health and safety regulations.

In your premises, *you are responsible for maintenance of the whole workplace including its contents.* This includes its equipment, devices and systems; ventilation; temperature; lighting; cleanliness and waste materials; room dimensions and space; workstations and seating; conditions of floors and traffic routes; falls or falling objects; windows and transparent or translucent doors, gates and walls; skylights and ventilators; ability to clean windows, etc.; safety organisation, etc., of traffic routes, doors and gates; escalators and moving walkways; sanitary conveniences; washing facilities;

drinking water; accommodation for clothing; facilities for changing clothing; facilities for rest; and facilities for eating meals.

This is a daunting list of responsibilities, but most can be managed by that scarce commodity – common sense. The following need some further explanation for the unwary employer.

Temperature

Temperature is important and can become a problem if you don't know the regulations. During working hours, the minimum temperature inside your workplace must be "reasonable". What is "reasonable"? At least 16 degrees Celsius is considered reasonable unless much of the work involves severe physical effort, when the temperature should be at least 13 degrees Celsius. Make sure that heating or cooling systems do not cause an escape of fumes that are likely to be injurious or offensive to anyone. Curiously, there is no specified maximum temperature for a workplace. In any case, temperatures alone may not ensure reasonable comfort because of other factors such as air movement and relative humidity.

Working in cold or hot conditions

If the workplace is a cold store, the temperature will be below 13 degrees and obviously this is not a reasonable temperature as previously defined. To compensate you will have to provide warm, protective clothing for staff working in that environment.

At the other end of the scale, if the workplace is a foundry where there are extremely hot temperatures, although there is no defined maximum temperature as mentioned above, this would not normally be considered a "reasonable" temperature in which to work. Common sense and good management practice suggests that you should provide a supply of cold drinks and frequent rest breaks away from the heat.

Temperatures can be a touchy subject and under the Regulations you must provide thermometers to enable your staff to gauge the temperature in their workplaces.

Space

Space is another contentious subject amongst staff. Overcrowding and cramped conditions are an obvious risk to health and safety and again common sense would suggest that they should be avoided.

You should be aware that there is a minimum amount of work-space catered for by law!

The minimum amount of cubic space allowed for each person in a workplace must not be less than 11 cubic metres. In calculating the cubic space for these purposes, you can disregard a ceiling higher than 4.2 metres (i.e. assume a height of 4.2 metres) and if a room contains a gallery or balcony, it can be treated as a separate room. If you visualise a workspace containing the average desk and chair 11 cubic metres is about right. A room 3.3 metres high containing a desk 2 metres in width and with a combined depth of desk plus chair as 1.67 metres, your figures would produce 11.02 cubic metres. Even the most pedantic HSE inspector would be happy with this, but in the real world allocation of office space is much more generous

Sanitary conveniences and washing facilities

You must provide suitable and sufficient sanitary conveniences and washing facilities for your staff. The rooms containing sanitary conveniences must be adequately ventilated and illuminated and must be kept clean and tidy. Ideally, you will have separate rooms for men and women except for washing of hands, forearms and face only. In small premises with few staff, it is permissible to have facilities for joint use for both female and male staff provided the convenience is in a separate room which can be locked from the inside. Washing facilities must include a supply of clean, hot and cold, running water, soap and towels (or the equivalent). There is a specified scale of toilets and washrooms which is roughly one of each for every 5 persons rising on a sliding scale to five of each for 100 persons.

Where facilities provided for your staff are also used by members of the public, you will need to increase the number of conveniences and washrooms necessary to ensure that staff can use washing facilities without undue delay.

Lighting

Your premises should be well illuminated by natural light so far as is reasonably practicable. This can be supplemented by artificial lighting where required. In addition, you need to provide emergency lighting in the event of failure of artificial light and where persons at work are specially exposed to danger.

Cleanliness and waste material

Your premises including furniture, furnishings and fittings must be kept reasonably clean. This includes floor, walls and ceiling surfaces. Waste material should be deposited in suitable receptacles and not allowed to accumulate unnecessarily.

Seating

Where staff perform work which can or must be done while sitting, you must provide them with suitable seating. Seats must be suitable for the people using them and the types of work they are carrying out. If necessary, suitable footrests must be provided.

Drinking water

Bottled drinking water has become standard fare and everyone seems to carry their own supply nowadays. However, you as an employer have to provide an adequate supply of wholesome drinking water for all persons at work. This can be by way of a drinking fountain or the bottled variety, or by the more conventional means of tapwater (provided the tapwater is fit for drinking).

Traffic routes

Pedestrian traffic

Slips, trips and falls are some of the main causes of accidents (and subsequent lawsuits) in the workplace. It makes sense to ensure that all stairs, steps and gangways are unobstructed. Floors and every traffic route should be properly constructed and free from holes, slopes or uneven and slippery surfaces. Staircases should have handrails.

Vehicle traffic

Where you have vehicles and work trucks circulating in and around your premises you need to route them in such a way that pedestrians and vehicles are separated as far as possible. All traffic routes should be suitably indicated. This can be done by painting yellow lines along the routes. Pedestrians and vehicles should be able to use a traffic route without causing danger to the health and safety of persons at work near the traffic route. The traffic route must be separated from doors or gates which lead on to it. Where vehicles and pedestrians use the same traffic route you must ensure there is sufficient separation between them.

Falls or falling objects

The regulations require that "suitable and effective measures shall be taken to prevent any person from falling a distance likely to cause personal injury and any person from being struck by a falling object likely to cause personal injury." This would include provision of handrails, appropriate channelling of overhead machinery, etc. Where there is a risk to health and safety on your premises you must clearly indicate the risk. If you have any tank, pit or structure containing a dangerous substance it must be securely covered (and labelled, although it doesn't actually say so in the Regulations) or fenced where there is a risk of a person falling into it.

Windows

Windows and transparent or translucent surfaces in doors or gates should be made of safety material or protected against breakage. Large glass surfaces which can be mistaken for open doorways should be appropriately marked to prevent accidents. Windows, skylights and ventilators should be capable of being opened without exposing any person to a risk of injury. When opened, they must not expose any person in the workplace to a risk to health or safety. All windows and skylights in a workplace must be designed or constructed for safe cleaning.

Doors and gates

Doors and gates should be suitably constructed and fitted with any necessary safety device. A sliding door or gate must have a device to prevent it coming off its track and any upward opening door or gate must have a device to prevent it falling back. A powered door or gate must have a protective mechanism to prevent it from trapping anyone.

Escalators and moving walkways

These should be equipped with any necessary safety devices and fitted with one or more emergency stop controls.

Clothes storage and changing facilities

Storage accommodation must be provided for staff's own clothing which is not worn during working hours and for special clothing worn at work but which is not taken home. You must provide suitable, separate facilities for staff to change clothing where they have to wear special clothing at work and cannot, for reasons of health or propriety, be expected to change in another room.

Facilities for rest and to eat meals

Where food eaten in the workplace could possibly become contaminated you must provide suitable facilities to eat meals. You must also provide suitable rest facilities for any pregnant woman or nursing mother.

Non-employees at work on your premises

You need to allow visiting employees the use of your facilities within reason, although the regulations are mainly concerned with safety rather than welfare.

Work Equipment

The Provision and Use of Work Equipment Regulations 1998 cover work equipment and their use and, as always, there are definitions requiring some explanation. "Work Equipment" is any machinery, appliance, apparatus, tool or installation for use at work. "Use" is any activity involving work equipment including starting, stopping, programming, setting, transporting, repairing, modifying, maintaining, servicing and cleaning.

Definition of work equipment

Work equipment includes:

- "tool box tools" – hammers, knives, handsaws, meat cleavers, etc.
- single machines such as drilling machines, circular saws, photocopiers, combine harvesters, dumper trucks, etc.
- apparatus such as laboratory apparatus (Bunsen burners, etc.)
- lifting equipment such as hoists, lift trucks, elevating work platforms, lifting slings, etc.
- other equipment such as ladders, pressure water cleaners, etc.
- an installation such as a series of machines connected together, for example, a paper-making line or enclosure for providing sound insulation or scaffolding or similar access equipment .

You can see from this that even the simple office contains work equipment which is subject to the Regulations.

Your work equipment must be suitably constructed or adapted for the purpose for which it is used or provided. In selecting work equipment, you must have regard to the working environment in which it is being used. Next, ensure that it is maintained in an efficient state, in efficient working order, in good repair. If it is machinery with a maintenance log, keep it up to date.

Where safety of work equipment depends on its installation, you must inspect it after installation and before putting it into service for the first time. Similarly, if it is assembled at a new site or in a new location

Where work equipment is used in conditions causing deterioration you must inspect it regularly and keep records of the inspections. *All work equipment, wherever used, should be accompanied by records of inspections.*

Repairs, modifications, maintenance or servicing of equipment can only be done by persons specifically authorised to carry out such work. You must ensure that such persons have received adequate training in repairing, modifying, maintaining and servicing work equipment

Ensure that all persons, including supervisors, who use work equipment have received adequate health and safety information and, where appropriate, written instructions relating to its use. The information and instructions required must be readily comprehensible to all concerned. As always, appropriate training must be given to all users and supervisors.

11

Control of Substances Hazardous to Health (COSHH)

These are covered by the *Control of Substances Hazardous to Health Regulations 2002*. Some years ago, research in America found that 30% of disease was occupationally based and recent UK statistics indicate that there are thousands of claims for disablement benefit for occupational diseases.

Going to work and being at work can be hazardous to health and you must ensure that your workers whilst at work are not exposed to any hazardous substances which your enterprise either uses or produces. Straight away you might say this is one area you don't need to worry about as you don't make or use anything of a hazardous nature but, in the world at large, there are four recognised health hazards – chemical, physical, biological and ergonomic. Can you honestly say that your staff will not encounter any of these at your place of business?

Let's just concentrate on chemical hazards. You might not make chemicals in your workplace but they are used in even the most innocuous office environments – from cleaning materials to printing inks and toners. Since 1988 we have had regulations concerning the use of some hazardous substances at work. These need to be assessed and *you* are responsible for the assessment.

Definitions

COSHH defines a "substance hazardous to health" as any substance identified as being **toxic, harmful, corrosive or irritant**; any

84

substance which is **carcinogenic;** any substance which has a **workplace exposure limit** defined in **EH40**; any micro-organism which is hazardous to health; and dust above a certain level of concentration in air. Here we are back to the gobbledegook, so what do these various words and phrases mean?

COSHH terminology

Term	Definition
Toxic	Pertaining to poison
Corrosive	A substance which will chemically attack materials or person damaging material and destroying living tissue.
Irritant	A substance or liquid which when in contact with the skin causes an inflammatory reaction such as dermatitis or when inhaled (e.g. dust) causes irritation or longer-lasting damage to the lungs (e.g. fibrosis).
Carcinogenic	A substance, exposure to which may cause cancer.
Workplace Exposure Limit	For a substance hazardous to health means the exposure limit approved by the Health and Safety Commission for that substance in relation to the specified reference period when calculated by a method approved by the Health and Safety Commission as contained in HSE publication "EH/40 Workplace Exposure Limits 2005" as updated from time to time.
EH40	EH40 is an HSE Guidance Note published annually containing the current statutory list of substances which have been assigned Workplace Exposure Limit.

You can find out if a substance has been defined as hazardous by referring to what are known as the CHIP Regulations – the *Chemicals (Hazard Information and Packaging for Supply) Regulations 2002* – which contain an approved list of substances, symbols which are printed on packages, descriptive risk phrases, and safety data sheets.

See Appendix 3 for additional details of COSHH, CHIP, and EH40.

Your main obligation under COSHH is to provide risk

85

assessments of any hazardous substances you may have in your workplace. Your first step is to identify the hazardous substance and provide answers to the following basic questions:

- What is it – dust, fumes, gas, etc.?
- Where is it?
- How much of it is present?
- What is it used for?
- Who and how many are exposed to it?
- How long is the exposure period?
- Are young persons or pregnant women exposed to it?
- Can other persons (not employees) be exposed to it?
- What control measures are provided – training, PPE, ventilation, respirator, etc.?
- Are the control measures used?
- Are the control measures effective?

You should then compare your answers with recognised standards (e.g. EH40) and advice (e.g. Codes of Practice) to arrive at an assessment of the likelihood of harm arising from the hazard and the severity of any consequences. As with all risk assessments they should be recorded, kept on your premises for availability to any visiting enforcement agency, and more importantly, copies provided for the information and training of your staff. A specimen COSHH assessment is shown on page 87.

Specimen COSHH Assessment

Does the workplace contain any hazardous substances? (If "No" no further action necessary)	Yes/No
List of substances:	
Hazards involved:	
Names of persons likely to be exposed to hazards:	
Evaluate risks to health and safety:	
Control measures:	

Assessor: ....

Signature: ....

Review date: ....

12

Display Screen Equipment (DSE)

Regulations governing the use of DSE are set out in the Health and Safety (Display Screen Equipment) Regulations 1992.

With the "computer age" now firmly established there can hardly be any business organisation which doesn't use a computer screen and keyboard of some sort. Regulations concerning the use of display screen equipment have been with us since 1992 but surprisingly are still not widely known in small businesses whereas large, unionised, businesses are all too familiar with them. They are important to the employer as incorrect use of keyboards can lead to hand, arm, and posture problems and the constant scanning of screens may cause eye problems. All of these can lead to time off work! You can dismiss the scare-mongering tales of the effects of radiation (from the screen) on pregnancies and eye cataracts which are without foundation. The real physical problems, however, can lead to claims for compensation which are to be avoided at all costs. It makes good sense then to get a working knowledge of the Regulations and what they involve.

First of all you need to know that the rules cover use *for or in connection with* work. You need to be aware that even when you allow staff to work at home, they are considered to be "at work". If you have supplied them with display screen equipment, the Regulations apply to them. In theory, if they are working at home you are also responsible for ascertaining the status of furniture and equipment at the user's home, even where that furniture and equipment is the property of the user.

When the Regulations first came into practice, portable systems, such as laptop computers, were excluded from the Regulations where use was intermittent and for short periods. These have now replaced many of the older desktop computers and where used must be taken into consideration for assessment purposes.

Users of DSE can be expensive so let's get hold of this term "user". A DSE user within the terms of the Regulations is an employee who habitually uses display screen equipment as a significant part of his normal work. But what is "a significant part of normal work"? Several factors must be considered here. For example, a part-time employee may use a display screen for an hour a day, which could be a significant proportion of his or her working time. On the other hand an hour a day for a full-time employee could be interpreted as insignificant in relation to time spent on other activities. Many of your staff will use computers in the course of their work without falling within the definition of user but where all or most of the following factors apply you can assume that your staff person is a "user" for the purpose of the DSE Regulations:

- The job requires the use of display screen equipment.
- The use of display screen equipment forms an important part of the worker's job description.
- The work often requires the use of a display screen for a period of an hour or longer.
- The display screen is used on most days, or every day.
- The worker has little or no discretion about the use of the display screen.

As already indicated users cost money in terms of eyesight testing and provision of spectacles and other remedies but where do those costs occur?

- **Expense number one:** Where a user requests an eye and/or an eyesight test, you must provide an appropriate test, carried out by a competent person, *at your expense.*

- **Expense number two:** Where the result of any eye and eyesight test indicates that corrective lenses are necessary, you must provide them *at your expense.*

This does not mean that all glasses must be supplied free of charge. If the employee already wears glasses for normal work, then there is no need for you to supply additional equipment, nor to pay for that which the employee already has. However, some spectacle wearers may require a different pair of glasses specifically for working with display screen equipment – in which case you would be responsible for their provision. Remember, you are only obliged to pay the minimum cost of obtaining suitable lenses and a standard frame. If the wearer wants more expensive designer frames, he is free to upgrade frames *at his additional expense.* Simply ascertain the cost of a standard pair of suitable glasses and offer to reimburse that sum, upon proof of purchase. If the glasses eventually supplied are not designed to correct vision at the normal viewing distances for display screen work, you are not obliged to contribute anything toward the cost.

If you have supplied spectacles and they are subsequently lost or accidentally damaged, you must arrange and pay for repair or replacement. *This may appear to be rather unfair but it is because, effectively, the specs are your property.* This is in line with rules governing the issue of personal protective equipment, the cost of which must be met by the employer. If you felt that an employee was not taking reasonable care of company property, you could invoke disciplinary procedures.

Display screen equipment is usually an integral part of a **workstation**. For these purposes, a "workstation" is an assembly consisting of display screen equipment and any peripheral accessories such as a disk drive, telephone, modem, printer, document holder, work chair, desk, or work surface and the immediate surrounding work environment.

As an employer you need to assess workstations provided for your staff to ensure that their use does not present any risk to their health and safety. *As with all assessments, you need to keep written records or keep them in some other retrievable format.* (Specimen assessments are shown at appendix 3.) You should also provide your staff with copies of their assessment to give them the opportunity to dispute any contentious points it may contain. Where risks are very low, or where the workstation is temporary, you don't need to formally record the assessment. Formal records however are essential for providing a defence in civil claims where negligence is alleged. Most employers use a self-assessment checklist and this can also be adapted for any "home-worker".

You must review any assessment when it is no longer valid, or if there has been a significant change in the lighting, the furniture, the task, the software, or a change of location.

Breaks from screen-viewing

Because of the potential stressful nature of working with display screens and keyboards, workers need to have periodic breaks. There is no stipulated maximum time for a period of work with display screen equipment. Neither is there a requirement for fixed breaks. You should structure the working day as far as possible to incorporate regular off-screen activities – ideally away from the screen. This doesn't mean enforced idleness. Each task should be evaluated to see where appropriate and natural breaks occur. Can you introduce some other activity to minimise the length of continuous on-screen activity?

Try to give your staff as much flexibility and control over the duration of screen work as possible. Frequency of breaks is more important than duration – three breaks of five minutes, spread evenly throughout a period of work, is considered more beneficial than one 15 minute break. Pressure of deadlines, high workloads, personal commitment and enthusiasm can increase physical and mental stress, and may result in breaks being ignored. You should encourage staff to adopt sensible work routines to avoid this.

Eyesight tests

Users are entitled, but have no obligation, to undergo an eye and eyesight test, at your expense.

An eyesight test can be performed by a competent person using an eye-screening device. You should inform employees that eyesight-testing is not a substitute for a full eye-test which is a test of visual capability over a range of viewing distances which can only be carried out by an optician. Obviously a full eye test is more expensive than an eye-sight test.

An eyesight test should assess visual capability at the normal viewing distance for a display screen. Where the result indicates that vision requires correction, the user should be advised to consult an ophthalmic optician for a full examination.

Where the user wishes for an eye-test, rather than an eyesight test, you must refer him to an ophthalmic optician or a medical practitioner. You can require the employee to be tested by an optician of your choice. This allows you to make an arrangement with, say, a local optician where all your staff will be tested to the same standard. The results of any sight-test may only be disclosed with the consent of the person tested.

*It is perfectly legitimate for you to require a **prospective** employee to make their own arrangements for eye-testing and to demonstrate that their vision is suitable for the position to which they seek to be appointed. You only become liable for costs after they have been appointed!*

A test requested by a new employee must be carried out before that employee commences work with display screen equipment. Similarly, where an existing employee's work changes to the extent that he becomes a user, a requested test must be carried out prior to the new duties being undertaken. You need to make provision for testing at reasonably regular intervals afterwards always remembering however that a user may request an eye test if he is experiencing eyestrain or other problems. You should be guided by the clinical judgement of the optometrist or medical practitioner who carried out the previous test.

Training

Training of staff is an on-going requirement in all areas but there is a specific requirement to train users in the health and safety aspects of using their workstation as well as the mechanics of the job. Training should include how to adjust furniture and equipment; how to arrange the workstation's layout to best effect; arrangements for and importance of taking breaks; eye and eyesight testing entitlement and arrangements; and the DSE Regulations and the user's role in meeting them.

You should keep a record of all training given. Users don't need to be trained typists nor do you need to train them in how to type. Your training programme should teach users to recognise and understand risks associated with using display screen equipment, the recognition of symptoms related to adverse health conditions, and the procedures for reporting problems to you. Where you alter a workstation substantially or introduce new software you must give additional, appropriate and adequate training.

You would be correct in assuming that this is an indigestible amount of bumf to get across to each new employee and user!

The simple answer to the problem lies in the production of a DSE policy which you can give to all members of staff or make available electronically on computer. The following précis puts the Regulations in a nutshell and the accompanying specimen assessment illustrates how simple and easy the whole process is.

Display Screen Equipment Regulations 1992

DSE is any alpha-numeric or graphic display screen. Window type-writers, calculators, cash-registers, and displays in drivers cabs or machine control cabs are excepted.

Regulations apply to "users" (and self-employed operators) e.g. word-processor operator, secretary, etc who:

- depend on DSE to do their work
- need significant training or special skills
- normally use for continuous periods of 1 hour or more
- are required to carry out fast transfer of information between themselves and the software package.

Regulations are designed to **prevent**:

- work-related upper limb disorders – neck, shoulders, arms, wrists

- eye strain
- stress

Regulations cover assessment for:

- users
  - provision of eye tests/spectacles
  - breaks/changes of activity
  - information and training
  - arm angle and wrist position

- workstations/environment
  - space requirements
  - lighting
  - noise
  - software
  - keyboard
  - chair adjustment (back/height)
  - reflection and glare
  - heat and humidity
  - screen desk/surface

Self-assessment checklist

DISPLAY SCREEN EQUIPMENT REGULATIONS 1992 SELF-ASSESSMENT CHECKLIST			
Name................................Department................................Date..............			
*Please complete the checklist as accurately as possible by crossing out the terms which **do not** apply. Please return to your manager as soon as possible.*			
Working environment Describe you workstation lighting level. Are there reflections on your screen? Can you control the lighting?	Just right Sometimes Yes	Too dark No	Too bright All the time Sometimes
Space Is there enough space around your workstation?		Yes	No
Temperature/humidity What is the usual temperature at your workstation?	Too hot	Too cold	Comfortable

Noise Does noise from other work equipment interfere with your concentration?	Yes	No	Sometimes
Furniture Is your seat height adjustable? Is your seat backrest adjustable for height and angle? Is your chair stable? Does it have castors? Is it in a good state of repair?		Yes Yes Yes Yes Yes	No No No No No
Desk Is the height of your desk suitable and comfortable? Does it have a non-reflective matt surface? Is the desk surface sufficiently large for all your peripheral equipment?		Yes Yes Yes	No No No
Document holder Do you use a document holder? Is it adjustable for your requirements? Do you know where to position it in relation to your screen?		Yes Yes Yes	No No No
Footrest Have you been supplied with a footrest, if one is required		Yes	No
Display screen equipment Is the screen free from glare and flicker? Can the screen be tilted and swiveled? Can you adjust for brightness and contrast? Is the screen height at a comfortable level? Is the keyboard separate from the monitor? Is the keyboard height adjustable? Can you rest your hands in front of the keyboard? Are key symbols easily readable?		Yes Yes Yes Yes Yes Yes Yes Yes	No No No No No No No No
Software Do you understand the software?		Yes	No
Training Have you received training in the use of your workstation and display screen equipment? Have you received training in the use of the software? Do you understand the company DSE policy in relation to eye and eyesight testing? Do you know the procedure if you encounter any problems in relation to DSE work?		Yes Yes Yes Yes	No No No No

Required features of workstation and posture

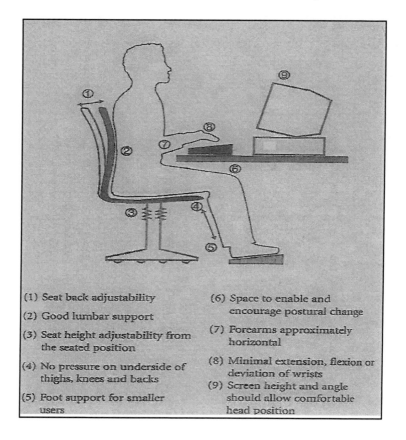

(1) Seat back adjustability

(2) Good lumbar support

(3) Seat height adjustability from the seated position

(4) No pressure on underside of thighs, knees and backs

(5) Foot support for smaller users

(6) Space to enable and encourage postural change

(7) Forearms approximately horizontal

(8) Minimal extension, flexion or deviation of wrists

(9) Screen height and angle should allow comfortable head position

13

Manual Handling

The rules governing manual handling are set out in the *Manual Handling Operations Regulations 1992* (MHOR).

Many employers, especially in office environments, take the view, wrongly, that their operation doesn't involve manual handling so there is no need to train staff in handling techniques. Do your employees lift, put down, push, pull, carry or move any movable objects in the course of their work? If so, you need give some thought to manual handling. Statistics show that 63% of manual handling injuries are simple sprains and strains, and you don't need to be involved in strenuous manual labour to develop a strain! Again, recent statistics show that over 600,000 people in UK have a health problem arising from manual handling at work and more than one third of accidents reported under RIDDOR result from manual handling.

In 1998 a nurse was awarded £78,000 from her employers who had asked her to move a locker some 600 yards across a main road. Why should a nurse have to move a locker? Her injury actually occurred in 1994, but this was more than a year after the MHOR were introduced. Obviously her employers were not complying with the Regulations. Injuries associated with manual handling are many and varied, from cuts, bruises, crush injuries to fingers, hands, forearms, ankles and feet, to muscle and ligament tears, hernias, knee, ankle and shoulder injuries and the universally popular slipped disc (prolapsed inter-vertebral disc).

Early factory regulations tried to avoid injury to workers by setting limits on the weight of any materials to be lifted. Different categories of weight were designated for men, women and young persons. This was a particularly stupid arrangement as it took no account of the individual physique, the weight, size and shape of the load and the working environment. The modern approach is to apply ergonomic techniques to the lifting and moving of objects with a view to preventing manual handling injuries. As a caring employer you mustn't put your employees at risk of injury from the manual handling of loads at work. So you'd better have an idea of what constitutes manual handling

Definition of manual handling

Manual handling is defined in the Manual Handling Operations Regulations as follows: *"Manual handling means any transporting or supporting of a load (including the lifting, putting down, pushing, pulling, carrying or moving thereof) by hand or by bodily force."*

For these purposes, a **"load"** is any movable object including people and animals.

Injuries covered by the regulations are any injuries arising from the manual handling activity.

Reducing the risk of injury

You need to take appropriate measures, in particular mechanisation, to avoid manual handling of loads. In other words, only move things by hand where they cannot be handled by mechanical equipment. If you cannot avoid manual handling, you must use the appropriate means to reduce the risk involved.

This is a simple two-step procedure:

- reduce the need for manual handling to essentials
- reduce any remaining risks involved in the handling through training, organisational measures, and improved workplace conditions.

Assessing the risk

You need to look at the type of work involved and assess any relevant health and safety condition, having particular regard to:

- **The characteristics of loads**
  Weight, size, shape, is it hot or cold, its centre of gravity etc

- **Worker information**
  Physical suitability to carry out the task in question

- **Personal Protective Equipment**
  The wearing of suitable clothing, footwear or other personal effects

- **Training**
  Worker's possession of adequate or appropriate knowledge or training

You will find guidance in the Regulations on how to make assessments. They provide four factors for consideration: the task, the load, the working environment and individual capacity. Each factor poses a number of questions which in turn give rise to some common-sense solutions.

Tasks

Does the task involve holding or manipulating loads at a distance from the trunk? Is it necessary to twist the trunk, stoop, or reach upwards to lift or move the load? Does handling involve any lifting, lowering, carrying distances, pushing or pulling, to excess? Does handling involve frequent or prolonged physical effort? Does the task involve a rate of work imposed by a process? In World War II convoys, the speed of the whole convoy was geared to that of the slowest ship. Similarly, in a factory conveyor-belt process the rate should be geared to the pace of your slowest operator. Workers who are unable to keep pace are subjected to intense pressure and the risk of physical and mental injury.

There are a number of things you can do to improve the task layout. For example, if the best position for the storage of loads is waist-high, avoid high or low shelf positions for heavier loads. Can you introduce training in correct lifting techniques?

Can you improve the work routine by varying jobs, rest pauses, etc.? Can you introduce team handling? Consider using two or more to lift and use handling aids, e.g. slings or stretchers. You need to assess the use of personal protective equipment, which may be essential to use but can cause restriction of movement. Maintain your handling equipment and ensure easy access to it all times for your staff. Keep the safety of all machinery under review and ensure it is designed for safe handling.

Loads

Are they heavy, bulky or unwieldy, difficult to grasp, unstable? Are the contents likely to shift suddenly? Does the load have sharp edges? Is the load hot or otherwise potentially damaging? Those are the questions. The obvious answers are: make them lighter (smaller packages are easier to handle, but don't ignore the consequent increase in handling frequency); make them easier to grasp, e.g. handles, grips, indents; make them more stable, e.g. use slings for non-rigid packages; make them less damaging to hold, e.g. insulate, no jagged edges.

The working environment

Are there space constraints preventing good posture? Are the floors uneven, slippery or unstable? Are there variations in level of floors or work surfaces? Is the working environment subject to extremes of temperature or humidity? Are there ventilation problems or gusts of wind? Are there satisfactory or poor lighting conditions? Again the answers are obvious: remove any space constraints; improve the condition and nature of floors, e.g. clear up spillages, repair holes; avoid working at different height levels; review the thermal environment avoiding extremes of temperature and humidity; finally, check the lighting to ensure that it is adequate and efficient.

Individual capability

Does the job require unusual strength, height, etc.? Does it create a hazard to those who might reasonably be considered to be pregnant or have a health problem? Does it require special information or training for its safe performance? Consider the individual's physical capacity and medical condition giving special consideration to a worker who is disabled. You must make allowances for the condition of pregnant women and women returning to work after maternity leave.

Another factor to be taken into account is the involvement of protective equipment (PPE) where movement or posture might be hindered by its use. The HSE offer the following guidance on handling techniques which is aimed at avoiding injury:

- Plan the lift, asking whether help or handling aids are required.
- Place the feet apart to give a balanced and stable base for lifting.
- Adopt a good posture with the knees bent and the hands level with the waist when grasping the load, the back straight, maintaining its natural curve and with the chin tucked in.
- Get a firm grip with the arms within the boundary formed by the legs.
- Carry out the lifting movement smoothly, raising the chin as the lift begins and keeping control of the load.
- Move the feet rather than twisting the trunk when turning to the side.
- Keep close to the load for as long as possible, keeping the heaviest side of the load next to the trunk.
- Put down the load, then adjust the positioning afterwards.

The following flowchart gives an indication as to how to carry out a manual handling assessment:

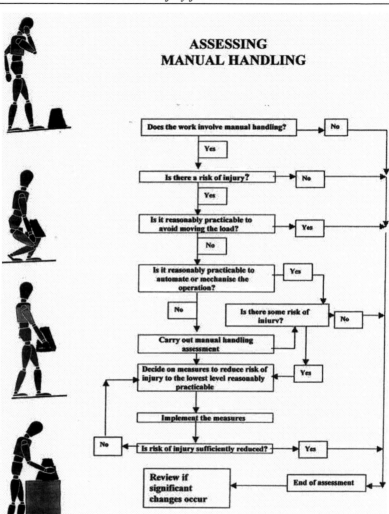

ASSESSING MANUAL HANDLING

Does the work involve manual handling? → No

Yes ↓

Is there a risk of injury? → No

Yes ↓

Is it reasonably practicable to avoid moving the load? → Yes

No ↓

Is it reasonably practicable to automate or mechanise the operation? → Yes

No ↓

Is there some risk of injury? → No

Carry out manual handling assessment

Yes

Decide on measures to reduce risk of injury to the lowest level reasonably practicable → Yes

Implement the measures

No ← Is risk of injury sufficiently reduced? → Yes

Review if significant changes occur ← End of assessment

102

Manual handling assessment checklist

<table>
<tr><td colspan="2">

MANUAL HANDLING OF LOADS
CHECKLIST

Note: This checklist may be copied freely. It will remind you of the main points to think about while you:
– consider the risk of injury from manual handling operations
– identify steps that can remove or reduce the risk
– decide your priorities for action

</td></tr>
<tr><td colspan="2">

Summary of assessment

</td></tr>
<tr><td>

Operations covered by this assessment:

Locations:
Personnel involved:
Date of assessment:

</td><td>

Overall priority for remedial action: Nil/Low/Med/High
Remedial action to be taken:

Date by which action is to be taken:
Date for reassessment:
Assessor's name

</td></tr>
</table>

Section A – Preliminary

	Yes/No
Q1. Do the operations involve a significant risk of injury? *If 'Yes' go to Q2. If 'No', the assessment need go no further. If in doubt, answer 'Yes'.*	
Q2. Can the operations be avoided/mechanized/automated at reasonable cost? *If 'No' go to Q3. If 'Yes' proceed with these actions and then check that the result is satisfactory.*	
Q3. Are the operations clearly within the company's current safety guidelines? *If 'No' go to Section B. If 'Yes' you may go straight to Section C if you wish.*	

Section B – More detailed assessment, where necessary
If the answer to a question is 'Yes' place a tick against it and then consider the level of risk.

	Yes/No	Level or risk (low/med/high)	Possible remedial action (make rough notes in this column in preparation for Section D)
The tasks – do they involve: • holding loads away from the trunk? • twisting? • stooping? • reaching upwards? • large vertical movement? • long carrying distances? • strenuous pushing or pulling? • unpredictable movement of loads? • repetitive handling? • insufficient rest or recovery? • a work-rate imposed by a process?			
The loads – are they: • heavy? • bulky/unwieldy? • difficult to grasp • unstable/unpredictable? • Intrinsically harmful (e.g. sharp, hot)?			
The working environment – are there: • constraints on posture? • poor floors? • variations in levels? • hot/cold/humid conditions? • strong air movement? • poor lighting conditions?			

(Section B contd.)			
Individual capability – does the job: • require unusual capability? • hazard those with a health problem? • hazard those who are pregnant? • call for special information/training? **Other factors –** Is movement or posture hindered by clothing or personal protective equipment?			

Section C – Overall assessment of risk

	Insignificant/Low/Medium/High
Q. What is your overall assessment of the risk of injury? *If 'Insignificant' the assessment need go no further. Otherwise, go to Section D.*	

Section D – Remedial action

What remedial steps should be taken, in order of priority?
1. ....
2. ....
3. ....
4. ....
5. ....

And finally

Complete the summary at the top of the form.
Compare it with your other manual handling assessments.
Decide your priorities for action.
TAKE ACTION.............AND CHECK THAT IT HAS THE DESIRED EFFECT.

14

Employee Information

There are a number of requirements regarding information to be given to employees. To start with, you need to provide your employees with summary information about the current health and safety legislation together with the addresses and telephone numbers of the health and safety enforcing authority and the Employment Medical and Advisory service for your area.

You can do this either by giving each employee a slip containing the information or by purchasing what is known as the approved poster **"Health and Safety Law – what you should know"** from HSE, HMSO or larger bookshops. This a large laminated poster containing basic health and safety information and blank panels in which you must insert the addresses and telephone numbers of your enforcing authority and EMAS office and the names of any employee or management representatives. You must then display the poster in an accessible place where the employee can see it.

If you decide to give the information individually you can provide a leaflet obtainable from the HSE entitled **"Health and Safety Regulation – a Short Guide".**

Under the Management Regulations you must also give your employees understandable information on:

- risk assessments
- the risks identified by the assessments
- the preventive and protective measures
- danger areas and procedures concerning serious and imminent danger
- emergency procedures

- names of competent persons responsible for implementing emergency procedures
- notice of any risks provided by other "sharing" employers .

The term "understandable" information is important because of your responsibility regarding any employee perhaps with learning difficulties or of ethnic origin who may have language difficulties. You may have to make special arrangements for special training, translation or making use of diagrams or symbols. Safety signs are an obvious way of communicating health and safety information and it will come as no surprise that they are subject to regulations and a British Standard. As well as conventional signs the regulations cover acoustic signals, illuminated signs, hand signals, and signs for internal traffic routes. If your risk assessments show that a risk cannot be avoided or minimised except by providing warning signs then appropriate signs must be provided.

You must also provide employees with information concerning the location of the Accident Book; First aid Box; the names and locations of your first aiders, if any; and where they can access a copy of the health and safety policy document.

$$15$$

Training

Recruiting personnel who are already fully trained obviously gives you a head start with your legal obligation to provide competent staff. However, you are still legally obliged to give staff proper training. You must provide your employees with the information, instruction, training and supervision necessary to ensure, so far as is reasonably practicable, their health and safety at work. Training must start when they join your organisation, when you transfer them or give them new responsibilities, or when you introduce new work equipment or a new system of work – and you need to give special attention to the needs of young workers.

Induction training

When you take on new staff you should provide them with induction training. This is important both for you and the trainee. It provides the necessary information for his personal safety and that of his co-workers, and familiarises him with your safety culture and attitude towards safety

A simple way to do induction training is to use an induction checklist which incorporates all the information concerning hazards and safety features which you need to impart. Get the trainee to sign the completed checklist, and retain a copy in his personnel file. In any subsequent investigation into an incident or accident you have documentary evidence that you provided basic training to the employee when he joined your organisation.

A simple checklist might include the following information:

- the hazards and risks in the workplace
- the precautions and protective measures required
- the fire and emergency procedures
- the first aid arrangements
- where to obtain help or advice
- access limitations, no-go areas
- key members of staff, supervisors, first aiders, fire marshals, etc.

Follow-up training

The remainder of the training requirements under the Management Regulations – relating to transfer, new technology and new systems of work – should be addressed by way of a **training needs analysis** for the individuals concerned. "Training needs analysis" is one of those jargon phrases found in management textbooks. There is no mystery about it and it is a relatively simple operation based on the following three stages:

1. Analyse the hazard/safety factors of the particular job.
2. Decide the standard of skill, knowledge and experience required to perform the job safely.
3. Assess the individual's existing abilities, expertise and competence.

The shortfall between 2. and 3. is the training need which you should address. *All such training must take place during working hours.*

The following case underlines the importance of training. A cleaning contractor, employed to clean vehicles at a site, terminated its agreement with a company. A company driver offered to clean the vehicles for paid overtime using a drum of cleaning chemical which was in store. The chemical in question had been overlooked when a chemical risk assessment was carried out. The driver was not told to take any precautions or given any instruction on how to use the cleaning fluid. After using the chemical, he suffered burns to his hands and ankle and had to attend hospital for treatment. The company had failed on two counts – no risk assessment and failure to give instructions or training – and was duly prosecuted for both.

16

Disability Discrimination

The Disability Discrimination Act 1995 made it an offence to discriminate against a disabled person. Initially, small businesses (less than 15 employees) were exempt but amending regulations in 2004, 2005 and 2006 brought small employers, together with a number of excluded occupations such as police, fire brigade, prison service and partners in business partnerships, within the scope of the legislation.

For the purposes of these rules, a disabled person is defined as any person suffering from a physical or mental impairment which has a substantial and long-term adverse effect on his ability to carry out day-to-day activities. "Impairment" is any condition affecting mobility, manual dexterity, physical co-ordination, continence, lifting or carrying ability, speech, hearing or eyesight, memory or ability to concentrate, learn or understand, and perception of the risk of physical danger.

The 2005 Act, already mentioned, amended the definition of disability in respect of people with mental illness and declared people with HIV infection, multiple sclerosis and cancer to be disabled for the purposes of the Act.

The anti-discrimination rules

Discrimination occurs where you, as an employer, for a reason relating to a employee's disability, treat him less favourably than you would treat other able-bodied employees unless you can show that your treatment of him was justified. Discrimination has to be avoided in every aspect of employment from dealing with recruitment advertising, application forms, interviews, proficiency testing, job offers, terms of employment, promotion, transfer or training opportunities, work-related benefits, and dismissal or redundancy.

You also have a duty to consider making reasonable adjustments to those employment arrangements or physical features of your workplace which might put a disabled employee at a disadvantage. These adjustments might include some of the following:

- reducing the amount of work allocated to the disabled person
- transferring him to a different job
- introducing more flexible hours
- giving him additional training
- modifying his equipment
- making instructions and manuals more accessible
- providing a reader or interpreter
- making adjustments to the building.

As well as practicability and reasonableness, adjustments to premises are affected by leases and building regulations. An employee who complains of discrimination can take his complaint to an employment tribunal if it can't be resolved internally with his employer. For your part as an employer, there is an informative Code of Practice available on the interpretation and application of the Act, which can be purchased from HMSO.

17

A Safe Place

Your premises have to be safe, both for your employees and for any visitors. Before addressing the problem of dealing with visitors, etc., you first of all have to make sure your place of business is a safe place for your own staff. To ensure, so far as is reasonably practicable, the health, safety and welfare at work of all employees, you have to provide a place in which they can work safely; then you must provide safe systems of working; and follow this up by employing competent people who are well supervised. There is little point in having a safe workplace and good working practices if you employ untrained, unsupervised and irresponsible workers.

Prior to the HSWA, employers of factory workers were legally obliged to ensure that the place of work (factory) was safe, first of all, with regard to its structure, and then with regard to the provisions of health, safety, welfare and hygiene for the employees. If you were not a factory worker your employer was not subject to the law!

The HSWA recognised this lapse and extended these requirements in general terms **to all employers**. The situation now is that you, as an employer, must ensure, so far as is reasonably practical, that any workplace under your control is kept in a safe condition and does not expose any employee to a health risk. This applies to buildings in relation to the building structure (walls, ceilings and floors) which must provide a reasonable standard of care. Floors must be sound, lifts must be in good working-order and properly maintained, staircases must be well-lit, and passages and roadways must be kept free from obstruction. The safety requirement also applies to open-air sites, boats, aircraft, tents and temporary structures such as scaffolding.

Getting to and from workplaces must also be safe and you must remember that access/egress conditions apply whether the workplace is a forklift truck, a scaffolding platform, or a confined space. Workplace safety is specifically dealt with in the *Workplace (Health, Safety and Welfare) Regulations 1992* but the following regulations are also relevant:

- *The Management of Health and Safety at Work Regulations 1992*
- *The Provision and Use of Work Equipment Regulations 1992*
- *Personal Protective Equipment Regulations 1992*
- *The Construction (Design and Management) Regulations 1994*
- *The Construction (Health, Safety and Welfare) Regulations 1996*

Both sets of Construction Regulations are only relevant to the construction industry so if your organisation is office-based you can safely ignore these two.

Look at your place of work objectively. Decide whether or not any breaches of legislation are taking place. Ask yourself the following questions:

Safety checklist

- Are the entrance/exit doors to your premises well maintained?
- Are the corridors and stairs well carpeted or have they a non-slip surface?
- Is the staircase fitted with handrails and is it adequately illuminated?
- Is the lift, if you have one, in good mechanical, working condition?
- Is the office itself properly furnished with good lighting?
- If it has central heating and/or air conditioning, are they efficient?
- Is there adequate space for every employee?
- Is there an efficient alarm system?
- Are the means of escape from the building well lit and easy to access?
- Are fire-exits well signed?
- Do panic-bars and escape mechanisms on doors work efficiently?

This is by no means an exhaustive list but if you can say "yes" to these you can assume that you are within the law and are providing a reasonable standard of care. However, it doesn't follow necessarily that where some negative responses occur you are in breach of the law. For example, consider this scenario: a female employee, on entering your premises, slips on the staircase, falls and breaks an arm. The staircase is brightly illuminated and has a non-slip surface but some food has been spilled on the stairs making them slippery. She was wearing high-heeled shoes. She was carrying an armful of files and was hurrying. There were no witnesses to the accident. The accident occurred just prior to the daily maintenance inspection of the stairs when warning signs pending repairs or cleaning would have been placed and there is no evidence to show that the food spillage was of long standing. Have you breached the duty of care in these circumstances? Remember, the use of the phrase "so far as is reasonably practicable" indicates that breaches of the Act are not absolute but qualified and the reasonable practicability of your arrangements for the maintenance of the staircase must be examined before apportioning blame. Your stairs have a non-slip surface and are well lit; you have the stairs inspected daily; you exhibit signs when repairs or cleaning are being carried out; and the food spillage may have happened just prior to the accident. Under the circumstances it could be argued that you had done everything reasonably practicable to ensure the safety of employees, and others, who might use the staircase. What about the employee? She was hurrying – carrying files – wearing high-heeled shoes – could these be contributory factors? They may or may not be, but the example shows that apparent health and safety breaches are not always as straightforward as they might seem. Providing a safe place of work is a matter of ordinary sound common sense but a safe place can always be compromised by the introduction of other factors such as systems and people.

18

Safe Systems of Work

A safe system of work refers to the way in which work is organised. This includes the layout of the workplace, the order in which tasks are carried out, and any special instructions or precautions in relation to any specific hazardous operation. For example, if your business involves the use of machinery, you must make sure that your machine operator has been properly trained in its use. He must know how to operate the machine safely without risk to himself or anyone else. If he follows your instructions governing the safe use of the machine, observes all relevant safety precautions, wears or uses the appropriate protective clothing or equipment, does not permit any unauthorised person to use or interfere with the machine, and operates it at the correct speed, you have provided a safe system of work for your employee.

Providing a safe system of work however is only half of the picture. You must also take reasonable steps to ensure that it is put into practice. If a risk assessment has shown there is a need for protective equipment, you must not only supply the equipment but ensure, by effective supervision, that it is being properly used.

Safe systems are a combination of factors – the systems themselves, how they are applied, how they interact, workers, machinery, hardware, software, instructions, protective equipment, manning levels, supervision, safety audits, safety inspections, safety policies, risk assessments, safety method statements, permits-to-work, induction training, tool-box talks – the list is endless.

The failure to provide a safe system of work has been illustrated in case law. In one important case, a social services officer was employed in an area with a high proportion of child-care problems. He suffered a nervous breakdown because of stress resulting from a high workload. On recovering, and before resuming work, his employers agreed he would receive assistance in order to lessen his burden. However, he was not given sufficient assistance and six months later suffered a second breakdown which forced him to give up work permanently. He sued his employer and the judge in the case decided in his favour saying that it was part of the employer's duty of care to provide a safe system of work. By failing to provide the necessary assistance to reduce the pressure from his workload, it was reasonably foreseeable that further psychiatric damage could occur. *(Walker v Northumberland County Council 1994)*

Competent and safe fellow-workers

Competence in this context has a general meaning relating to the common sense, experience and training of employees. This commences with the appointment of staff and, whilst bearing in mind discrimination law, it is probable that you will appoint people who are either already skilled in the job you want them to do or have the aptitude to acquire those skills after *training*. Where work is hazardous, you should not involve staff in the work until they have been given appropriate training and instruction. Similarly, where you introduce new methods of work or new technology, staff should not be involved until suitably trained.

Young employees

As discussed above, you need to be even more careful when taking on youngsters under the age of 18. Employ the ubiquitous risk assessment and *provide details of the assessment to the young person's parents or guardians.*

115

19

Visitors, Contractors and Customers

Having established your obligations to your employees you can then determine your requirements in relation to *other persons*. Who are these other persons? Quite simply they are persons, not in your employment, who might be affected by your business. You must ensure, so far as is reasonably practicable, that they are not exposed to risks to their health and safety by the way in which you conduct your business. Other persons, then, are **visitors** to your premises, including contractors you do business with, members of the public, customers, occupiers of neighbouring premises, children and even trespassers. They also include **contractors** like the electricians, plumbers or other traders who come into your premises to install, service or repair your fittings or equipment. Your duty of care extends to these people (though of course this is a reciprocal duty).

Remember that the self-employed are also bound by this obligation.

You need to carry out a risk assessment to ensure that employees of other employers and members of the public, are not affected by your operations. This is of particular importance in the construction industry where you may be occupying the same site with many other employers and also in large commercial premises, again, where you may occupy one of several tenancies in the same premises. The HSWA recognises that in these circumstances there may be more than one person exercising a degree of control. For example, in a multi-storey building containing companies or organisations which rent space from a landlord, each individual company is responsible for control over their part of the premises and must maintain good access and egress to and from it. A shopping mall is a good example of this where operators of individual shops and boutiques are responsible for the actual premises they occupy but the landlord is responsible for maintaining access and egress to the common areas of the premises, lift lobbies, external fire-exits, and plant rooms. Briefly, anyone who has partial control of premises must ensure the safety of any person who enters their premises to carry out work.

If you are in shared occupation with other businesses, you must exchange relevant health and safety information with those other employing organisations.

As an employer, or if you are self-employed, you must give any relevant safety information to non-employees who might be affected by your business. This is simply an extension of your duty of care towards the public at large and to contractors and their employees. It would cover such eventualities as fire, explosion, falls from unsafe scaffolding, and release into the atmosphere of harmful emissions. This is not as complicated as it sounds. Most of the information can be passed on by displaying appropriate warning notices on your premises. It makes sense that the protection you provide to employees should equally be made available to visitors, members of the public and other contractors. However, the standard may need to be higher in the case of certain members of the public whose appreciation of the potential danger is limited, e.g. the disabled and the very young.

One example of an offence affecting "other persons" is the **emission of noxious or offensive substances** being emitted into the atmosphere from your premises (or into the ground and the underlying water table).

The following are actual examples of prosecutions for offences under HSWA:

- Scaffolding material fell from a building site in central London onto two cars below because a sub-contractor had overloaded a scaffold. The prosecution said that the accident happened because the scaffolding company was stacking dismantled material onto the top platform of the scaffold, which gave way under the weight. It was alleged that the principal contractor was not sufficiently controlling the operation while the scaffolding company was not sufficiently supervising its employees. Both were found guilty, fined and ordered to pay costs.

- A truck driver was electrocuted at a construction site, when his lorry struck an over-head power cable. When he arrived at the site there was no one present from the company to tell him where to place his load. He therefore decided to unload next to an unfinished part on an all-weather surface. As he moved his tipper truck forward, the hydraulic ram touched an 11,000 volt overhead power line and he was electrocuted. No information about the risks of overhead power lines had been given because it was considered that there were no safety risks. The site owner was found guilty of failing to ensure the safety of a person not in his employment and failing to carry out a risk assessment. The company was fined and ordered to pay costs.

- 20-metre scaffolding poles were being lifted by a crane from a lorry parked beside a construction site. The poles were being lifted using a single chain. As a bundle of 50 poles was being lifted, they slipped from the chain and fell down into the busy street below. A pedestrian was struck by a bouncing pole, suffering cuts and bruises and later suffered panic attacks. Two other poles pierced the bonnet and windscreen of a taxi. No method statement for repetitive lifts had been prepared by the company and there was no three-way radio link between the slinger, loading the poles from the lorry, the crane operator, and the scaffolder receiving the poles. Instead, the scaffolder relied on hand signals from the crane operator. The trigger of the accident was failure by the slinger to follow the correct procedure in which he had been trained. The cause of the damage was primarily the failure to secure the load correctly as it should have been slung by using two chains. Another complaint related to signalling and communications. The company was fined for failing to secure a load properly and for failing to ensure, so far as is reasonably practicable, the health and safety of people not in its employment and ordered to pay costs.

Although these cases relate to the construction industry they illustrate quite clearly the need to take care that your business activities do not cause injury or damage to other persons and property.

Responsibility to customers

Most businesses involve the design, manufacture or supply of something to be used by other persons. A motorcar or aircraft designer would soon go out of business if the car or aircraft proved to be unsafe. Testing, research and examination must be carried out to ensure that designs and products are safe to use and are without risks to health. This is covered in great detail in the HSWA and in the Consumer Protection Act 1987. Looking at the Act, you can see that responsibility for safety goes "right back to the drawing board". Where a designer is negligent in the design of a machine or other article for use at work, he will be liable at criminal law under the HSWA.

Manufacturers, importers and suppliers are in exactly the same position. They have a similar duty to ensure that the necessary research is carried out to discover any potential risk to health or safety arising from any manufactured article or substance and, so far as is reasonably practicable, eliminate or minimise it. It is possible under the Act to transfer liability by written undertaking. This simply means that the responsibility placed on a supplier or manufacturer, etc., to ensure that an article or substance is safe may be shifted to another person provided there is a clear written agreement to that effect. This covers situations where an article has been ordered to a customer's own specification or is to become a component part of another article. Installers and erectors of equipment and machinery, for use at work, must ensure, so far as is reasonably practicable, that health and safety hazards do not arise from the method of erection or installation.

119

20

Employees' Responsibilities

The definition of "employer" and "employee" for the purpose of health and safety legislation have already been established (see page 23). In this chapter, the terms "work" and "at work" also need explanation. "Work" means work as an employee or as a self-employed person. Employees are "at work" throughout the time when they are doing something which is part of their employment. This may include, for example, travelling between different sites or buildings. Self-employed persons are at work throughout the time they devote to work as a self-employed person.

People undergoing training are also deemed to be at work, and this includes people taking part in training provided under government and other schemes.

So far we have been looking at employers' obligations, but what about employees? The Act states that *it is an employee's duty, while at work, to take reasonable care of his own health and safety and the health and safety of other persons who may be affected by his acts or omissions.* Also, if the law imposes any duty or requirement on an employer, or any other person, his employee must co-operate, so far as is necessary, to enable the employer to carry out or comply with the law

*Notice here that both **acts** and **omissions** are included.*

If the law requires you to provide an employee with personal protective clothing or equipment you must provide it at your expense but, on his part, failure to wear protective clothing would render him liable to prosecution. Similarly, failure to check machinery, which he is responsible for checking, could lead to disciplinary action and possible prosecution.

Irresponsible behaviour

Skylarking, horseplay and taking unsafe shortcuts has in the past led to prosecution. Fooling around, however innocently, can lead to accidents. If an employee has a tendency towards dangerous practical jokes in the workplace, which is foreseeable, you have a duty to prevent such activities in order to protect your staff. Again there is case law on the subject. In one particular incident the employer was aware of an employee's tendency to indulge in horseplay at the expense of his colleagues and had frequently issued reprimands and warnings. As a result of a prank by the employee in question one of his fellow-workers sustained a broken wrist. At court, the judgement was that the employer must have known that eventually his employee's behaviour would result in injury. The employer failed to take action to put an end to that type of conduct, to see that it did not happen again, and if it did happen again to take the appropriate disciplinary action. The employer was therefore in breach of the duty to provide competent staff. *(Hudson v Ridge Manufacturing Co. Ltd 1957)*

Effective supervision

Effective Supervision is an important factor in keeping the workplace safe and particularly in relation to the employment of young persons. A good supervisor keeps staff on their toes in respect of all aspects of the business including health and safety. The HSWA makes it an offence for any person, intentionally or recklessly, to interfere with or misuse anything provided in the interests of health, safety or welfare. The wording implies that the duty is not limited to employees. Any person so interfering is liable under the Act. Interference with fire-fighting or first aid equipment would fall into this category.

21

Managing Health and Safety

The management of health and safety is governed by the *Management of Health and Safety at Work Regulations*, first introduced in 1992 and updated in 1999 (and referred to in this book as the Management Regulations). Before looking at these specifically, it is as well to remember that the basis of all good management is simple:

- Plan what you want to achieve.
- Marshal and use your resources as effectively as possible.
- Control your operation closely.
- Keep a watchful eye on day-to-day events.
- Review your results daily, weekly, monthly.

The HSE use the flowchart set out on page 123 to demonstrate the above principles.

The final stage of this flowchart, "reviewing", is important because it highlights good and bad practices which you can eliminate or reinforce quickly for maximum benefit.

Again, it should come as no surprise to learn that under the Management Regulations, the safety perspective is to **plan, organise, control, monitor and review**. These methods echo the principles set out in the HSWA. If your enterprise is big enough you will have to form consultative mechanisms, e.g. health and safety committees etc, irrespective of trade union involvement.

The Management Regulations introduce risk assessments, health surveillance of employees, emergency procedures, provision of information and training for employees, so you need a working knowledge of these requirements.

HSE management flowchart

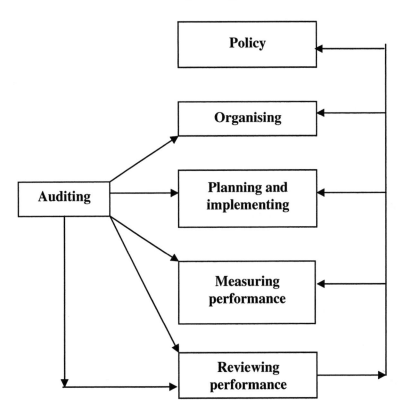

Health surveillance

If your business deals with the handling or making of hazardous substances, the Regulations require you to provide employees with appropriate health surveillance. The need for it should be identified by the nature of the job itself, e.g. working with lead, asbestos, certain solvents and chemicals etc., all of which require health surveillance as a matter of course, or by risk assessment. In the case of exposure to asbestos, employees must have a medical examination every two years and records of such examinations must be kept for at least 40 years due to the lengthy incubation period of asbestos-related disease. Naturally, if no such risks apply to your business, you can ignore this part of the Regulations.

Emergency procedures

These are straightforward enough. Principally they are your fire procedures and how to evacuate your premises in the event of a fire or other emergency. It is a simple matter to display the procedures by way of a "fire action" notice instructing employees and visitors what to do and where to go in the event of an emergency. Directional fire-exit signs must point people towards the nearest exit and these must inform by way of a **pictogram**, i.e. a sign which contains a symbol in addition to text. The "running man" symbol is a good example. You can obtain statutory notices for almost every conceivable emergency and fire situation from a number of reputable suppliers

Fire Action Notice

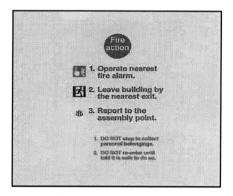

"Running man" sign

Under the Management Regulations, you need to deal with serious and imminent danger by setting out appropriate emergency procedures; nominating a sufficient number of **competent persons** to implement those procedures; restricting unauthorised access to dangerous areas; telling employees about any hazards and how to avoid them; enabling the employee to stop work immediately in an emergency and go to a place of safety; and preventing employees from resuming work until any danger has passed. Again in a larger organisation you will need to train and appoint persons to help with some of this but in a smaller one you can deal with it yourself and, of course, in the case of a sole trader you are the only candidate!

Shared premises

If you are sharing premises with other organisations you need to exchange health and safety information about risks and hazards which might affect your respective employees. If you are the main employer controlling a work site, you need to inform other employers of the site-wide health and safety arrangements and invite a sharing (reciprocation) of health and safety procedures, hazards and risks for the benefit of everyone on the site. Where there is no controlling employer, the employers concerned should appoint a health and safety co-ordinator.

Contractors

The situation where you are having work carried out on your premises under a service contract for cleaning, repair, maintenance, etc., or you are using "agency" employees to do temporary work for you is a case in point. The important principle here is that any persons who visit another employer's premises to carry out work must be provided with appropriate information and instructions regarding relevant risks to their health and safety. They in turn must inform you of anything of a risky or hazardous nature associated with their work.

Employees' duties

As has already been said, health and safety is not all one-way traffic, and under the HSWA employees must take reasonable care of their own health and safety and that of others who may be affected by their acts or omissions at work. Employees also have a duty *"not to interfere with or misuse anything provided by the employer in the interests of health and safety"*. The Management of Health and Safety at Work Regulations 1999 has reinforced this. When you provide your employees with any equipment, dangerous substance, transport equipment, means of production or safety device, you must give them any relevant training or instructions. Employees must then use such equipment etc only in accordance with the training and instructions you have given them. In addition, your employees must tell you, or their work colleagues, of any work situation which they reasonably consider to be a serious and immediate danger to health and safety. They must also inform you (and their colleagues) of anything which reasonably can be considered faulty in your health and safety arrangements.

Health and safety monitoring

One way of managing health and safety and incidentally helping to avoid accidents in your organisation is by constantly monitoring your safety performance. Your monitoring activities should be closely related to your **arrangements** for ensuring a safe and healthy workplace which are included in your Health and Safety Policy. Monitoring enables you to:

- analyse and evaluate certain health and safety tasks
- motivate staff, for instance, to use personal protective equipment,
- ensure that safe systems of work are being followed.

There are a number of ways in which you can do this.

126

Safety surveys

A safety survey is normally carried out by a specialist concentrating on specific aspects of safety. The survey is a very detailed examination of a critical area of operation or an in-depth study of the whole health and safety operation of your organisation. Areas covered might include health and safety management and administration, environmental factors, occupational health and hygiene provisions, accident prevention, and the current system for health and safety training of staff, from the Board of Directors to shop floor workers. Contractors and other groups who may enter the premises on a casual or infrequent basis might also be covered. The surveyor would then provide you with a report on an observation and recommendation basis. Recommendations would be based on the degree of risk, current legal requirements and the cost of eliminating or reducing the risks, e.g. the eventual replacement of old machinery and dilapidated buildings which may be totally unsuitable for modern work processes.

A survey conducted by a health and safety professional is expensive and smaller companies could probably monitor their health and safety requirements by using some of the following simpler techniques.

Safety audit

A safety audit is the measurement and validation of how your staff manage your health and safety programme compared with the instructions and standards you have laid down. You can do this yourself by looking critically at each area of your activities with the object of minimising injury and loss. Your starting point should be a full audit of your complete programme in order to identify its strengths and weaknesses. You can then make recommendations and set targets for any subsequent audit.

Safety inspection

A safety inspection is a scheduled inspection of a department or workplace. Usually this is done by your own staff. However, inspections are often carried out by external specialists, such as company health and safety specialists, plant safety officers, local management, trade union safety representatives, inspectors of the enforcing authorities, and insurance company surveyors. Examples of the types of inspection which might be carried out are as follows:

- **Enforcement agencies** carry out inspections to identify breaches of the law regarding the causes of a notified accident.
- **In-company health and safety specialists** inspect to identify hazards; to ensure that specific procedures designed to promote safe working are being operated; and to protect their employers from liability.
- **Safety representatives** carry out inspections to protect members of their trade union from hazards. Inspections focus on specifics such as maintenance standards, employee involvement and attitudes, working practices and housekeeping levels.

Safety tours

A safety tour is an unscheduled examination of a work area, carried out by a manager, possibly accompanied by health and safety committee members, to ensure, for example, that standards of housekeeping are at an acceptable level, fire protection measures are being observed and maintained, or personal protective equipment is being used correctly. (This is an extension of the "management by walking about" mentioned earlier.) Safety tours give an overall impression rather than a detailed analysis of hazards.

Safety sampling

This is measurement by random sampling where you can measure the accident potential in a specific workshop or process by identifying safety defects or omissions. You simply divide the area into sections and assign an observer to each section. Next, prescribe a route through the area and instruct the observers to follow the route within an agreed time-scale – say 15 minutes. Along the route, the observers note any specific safety defects (bad housekeeping, eye protection not being worn, incorrect handling procedures, etc.) on a safety-sampling sheet. Other aspects for observation may be included such as obstructed fire exits, lighting and ventilation problems, faulty hand tools and damaged guards to machinery. The observers should be trained in the technique and have a knowledge of the procedures and processes carried out. You then count the number of faults and non-compliances to provide results which you can collate and produce in graphical form.

Consultation with safety representatives

There has been a long-standing obligation for employers to consult union-appointed safety representatives under the *Safety Representatives and Safety Committees Regulations 1997*. This duty has now been extended by the *Consultation of Employees Regulations 2004* which require you to consult employees directly or through representatives of employee safety. Matters for consultation include:

- The introduction of any measure affecting the health and safety of the employees concerned.
- The appointment of persons nominated to provide health and safety assistance, and assist in emergency procedures
- Any health and safety training or information the employer is required to provide to the employees or the safety representatives.
- The health and safety consequences of the planning and introduction of new technologies in the workplace.
- The provision of any relevant information required to implement health and safety legislation.

If you use the employee representatives' route, you need to inform your staff of the names of those representatives and the groups of employees they represent (although they will already be aware!). They must be given all necessary information to enable them to perform their functions and participate in consultation. If you discontinue this form of consultation you must let your staff know.

Where consultation is direct, you must provide all the information to *all* your employees in order to let them participate fully in the consultations.

You are not obliged to disclose information that:

- does not relate to health and safety.
- is against the interests of national security.
- would contravene any prohibition imposed under any legislation.
- is related specifically to an individual (unless that individual has given his consent).
- would damage your business, or another person's.

Your employee representatives can approach you on any hazards, dangerous occurrences and general health and safety matters which may affect the health and safety of the staff they represent.

They can also approach any enforcing authority inspectors on their colleagues' behalf without your permission!

You must give them appropriate training and other relevant facilities to enable them to carry out their duties efficiently and all reasonable costs associated with the training, including travel and subsistence, are at your expense. Paid time off must be provided for candidates standing for election as employee representatives to allow them to perform their duties as candidates. Once elected, the representatives must again be given paid time off to perform their safety duties and to attend relevant training courses.

If employers refuse to allow employee representatives time off with pay to fulfil their duties, the representatives may complain to an employment tribunal.

The *Employment Rights Act 1996* has been amended to protect employees who participate in consultation with employers from suffering any detriment.

22

Penalties

The opening chapter of this book explained about registration with either the HSE or a local authority and how an authorized inspector can enter premises and take certain actions. If he finds evidence of a breach of health and safety law he may prosecute the offender. However, in many cases he will proceed by way of giving advice or issuing an enforcement notice. There are two types of notice:

- **Improvement notice** – this explains in writing what is wrong; what is required to put things right; and giving a minimum of 21 days in which to rectify matters.

- **Prohibition notice** – this is served by an inspector when he is of the opinion that an activity presents a risk of serious personal injury or danger. The notice must state the inspector's opinion; the nature of the risk; the specific law which is in breach or likely to be breached; and the notice must stop the specific activities until the situation has been remedied.

Failure to comply with an improvement notice may result in a prosecution at magistrates' court but it is possible to appeal to an employment tribunal within the 21 days. In such cases the notice is suspended pending the outcome of the hearing.

Failure to comply with a prohibition is much more serious and can attract a heavier penalty in court including imprisonment. Again, an appeal may be made to an employment tribunal but in this case the prohibition remains in effect until the tribunal has made a decision one way or the other.

Having bought this book and read it carefully you are unlikely knowingly to commit a health and safety offence. If you are unfortunate enough to transgress it would probably only be a minor transgression and probably dealt with by way of an improvement notice.

However, for the record it is worth remembering that breaches of health and safety law are criminal offences and if convicted can carry hefty fines and even imprisonment.

These are the offences and penalties under the *Health and Safety at Work etc. Act 1974:*

Offences

Summary offences (dealt with by Magistrates Courts)

- Contravening a requirement imposed by the HSC to order an investigation.
- Contravening a requirement imposed by an inspector.
- Preventing or attempting to prevent a person from appearing before an inspector, or from answering his questions.

Each way offences (dealt with by Magistrates or Crown Court)

- Failing to carry out one or more of the general duties of the 1974 Act.
- Intentionally or recklessly interfering with anything provided for safety.
- Levying payment for anything that an employer must by law provide in the interests of health and safety, e.g. protective clothing.
- Contravening any health and safety regulations.
- Intentionally obstructing an inspector.
- Intentionally or recklessly making false statements, where the statement is made to comply with a requirement to furnish information, or to obtain the issue of a document.
- Intentionally making a false entry in a register, book, etc. which is required to be kept.

- Falsely pretending to be an inspector.
- Contravening any requirement or prohibition imposed by an improvement notice or prohibition notice

Triable Only at Crown Court

- Doing an unlicensed act which requires a licence from the HSE.
- Attempting to acquire, possessing or using an explosive article or substance in contravention of the regulations.

Penalties

For conviction at a **Magistrates' Court** the maximum penalties are:

- £20,000 for breaches of Sections 2 to 6 of the 1974 Act.
- £20,000 and/or six months' imprisonment for breaches of improvement or prohibition notices.
- £5,000 for obstructing an inspector or other breaches of the remaining sections of the Act or dependent regulations

For conviction at **Crown Court,** the accused faces an *unlimited fine.* In the following cases there may also be *two years' imprisonment:*

- Certain offences involving required licences (stripping asbestos).
- Certain offences involving explosives.
- Contravention of an improvement or prohibition notice.

It is worth noting that, following non-compliance with an improvement notice or a prohibition order, where a person, after conviction, continues to contravene the improvement notice or prohibition order, he is liable to a maximum fine of £200 for every day that non-compliance continues.

Also, directors convicted of indictable offences relating to the management of a company may also be disqualified from holding a directorship for up to 5 years by a lower court and up to 15 years by a higher court.

23

Smoking

Cigarette smoking has always been a source of friction amongst employees. "No smoking" offices and factories led to smokers being allowed time away from desks and workplaces for 10-15 minutes (often twice per day) to satisfy their addiction. This in turn gave rise to annoyance amongst non-smokers who felt they were being unfairly treated in respect of "breaks". As from 1 July 2007, this petty annoyance finally disappeared as all enclosed public places and workplaces in England became "smoke-free". Similar regulations were already in place in Scotland, Wales and Northern Ireland. The legislation – the *Health Act 2006* – places a legal duty on occupiers or managers of premises to display no smoking signs on their premises and in their vehicles and as far as possible to enforce the ban. The legislation is welcome in that over 100,000 deaths from smoking occur annually in the UK. The law is designed to limit smoking in public places which might decrease the exposure of non-smokers to the effects of passive smoking. Difficulties may arise from the problems of enforcement as the employer or manager has the following legal obligations under the Act:

- The employer or manager must display appropriate no-smoking signs on the premises or in company vehicles.
- The maximum fine on conviction for an offence relating to the display of signs is currently £1,000 or £200 if dealt with by way of fixed penalty.
- A person who controls or manages smoke-free premises and fails to prevent smoking in the premises faces a maximum fine on conviction of (currently) £2,500. There is no fixed penalty procedure for this offence.

- An employee or visitor to the premises who ignores the signs or any instruction re non-smoking faces a maximum fine on conviction for smoking in a smoke free place of (currently) £200 or £50 if dealt with by way of fixed penalty.

Obviously employers and managers cannot supervise activities in company vehicles apart from ensuring that the appropriate signs are displayed. Here the responsibility for enforcement is delegated to the driver.

The fine details of the smoking regulations made under the Health Act are contained in the following:

- *Smoke-free (Premises and Enforcement) regulations 2006*
- *Smoke-free (Vehicle Operators and Penalty Notices) regulations 2007*
- *Smoke-free (Penalties and Discounted Amounts) Regulations 2007*
- *Smoke-free (Signs) Regulations 2007*

24

Waste and Pollution

All business organisations produce waste and yours will be no exception. If you are a small office-based company then you need not worry too much about it. Your small amount of waste will be covered by your local authority waste disposal arrangements.

However with a society that sees nothing wrong with putting spy-cameras in waste bins it is as well to try to keep ahead of the game. So what is waste? Simply, it is anything you discard – unwanted rubble, potato peelings, old newspapers, office files, old computer equipment – the list is endless. As a householder you will already be familiar with the emphasis on re-cycling and probably at home have different coloured bins for disposing of plastics, glass, newsprint and other household detritus.

Under normal circumstances, your business waste will be disposed of accordingly.

Hazardous waste

It becomes more complicated when your waste is classified as hazardous waste, i.e. waste which could be harmful to human health or the environment if it contains explosive, flammable, corrosive, oxidising or carcinogenic properties. All this is covered by the *Special Waste Regulations 1996* which contain a list of special wastes in its Schedule (a copy of which is shown at Appendix 6).

The following shorthand list gives some idea of what constitutes special waste:

- asbestos
- lead-acid batteries
- solvent-based inks
- chemical wastes
- fluorescent light tubes
- waste oil
- pesticides
- acids
- some prescription medicines
- electrical equipment containing cathode ray tubes.

If your business generates special waste, disposal is dealt with by a different and more complicated procedure involving a number of persons:

- the person getting rid of the waste – the consignor (you)
- the person transporting it to the waste disposal site – the carrier
- the person receiving the waste at the site – the consignee who is a waste site manager licensed to dispose of the waste, and
- the Environment Agency responsible for ensuring that the Regulations are complied with.

As the consignor you must store the waste in secure containers and inspect them weekly; keep an inventory; train your staff in how to handle the waste; use only registered or exempt waste carriers; use consignment notes and keep copies for three years; and register with the Environment Agency annually.

The disposal of the waste has to be documented by completing a five-part consignment note. Consignor, carrier and consignee all make entries in the consignment note and copies are kept by all the parties concerned. Don't worry too much about the other parties as long as you follow correctly your part of the procedure which is:

- complete parts A and B of all five copies of the note
- send one copy to the Agency
- complete part D of remaining four copies and keep one copy
- give the other three copies to the carrier
- file your copies of consignment notes on site and retain for three years.

This sounds like bureaucracy gone mad but has arisen as a result of the masses of waste generated by our consumer society coupled with the unscrupulous fly-tipping which is ruining our environment. Chemicals dumped indiscriminately can leach into the water table and sources of drinking water which is a highly dangerous and unsatisfactory situation.

At the beginning of this book the concept of the duty of care was discussed and it was suggested that this was the basis of all health and safety. There is a duty of care in waste management to ensure that your waste is handled and disposed of safely. The duty applies to anyone who produces, imports, carries, treats or disposes of controlled waste. There is no time limit to your duty of care. It is important to use registered waste carriers. If you are disposing of waste to a waste contractor, scrap metal dealer, or even your local council, make sure that they are authorised to deal with your particular type of waste. If an unauthorised carrier takes away your waste and fly-tips it, you can be held responsible for it.

If you are operating in a small office environment, you are not so likely to be involved with pollution problems. If you are in manufacturing or processing, your business may cause pollution by the impact it has on the environment in a number of ways:

- by burning of fossil fuels and materials creating dust, smoke and fumes
- oil spillages
- accidental release of raw sewage
- allowing chemicals, salt, wash waters, waste products, trade effluent, slurry and fuels to enter the soil, surface or ground water
- causing excessive noise, heat or vibration
- emissions from transport.

You should be aware that pollution covers emissions into the open air and water and onto or into land if they are harmful to humans, animals or plants, cause damage to the environment, cause a nuisance to the public or are unsightly.

The following processes require permission under *Integrated Pollution Control* or *Pollution Prevention and Control* legislation:

- combustion and incineration
- metal manufacture
- certain asbestos activities
- coating activities , printing and textile treatment
- treatment of animal and vegetable matter

You can obtain advice on *Integrated Pollution Control* and *Pollution Prevention Control* from your local authority or environmental regulator. A useful website is www.netregs.gov.uk.

25

Health and Safety:
A Potted History

Current safeguards against manual handling injuries would have been manna from heaven for those poor souls who built the Egyptian pyramids, the Mayan and Inca temples, and European medieval fortresses, churches and palaces but of course those monuments would never have been built! Early health and safety was influenced by the need to counteract the spread of plague and disease in communities and the introduction of sanitation, quarantine measures and controls on the sale of contaminated food were early milestones in the development of health laws.

The first real stirring of interest in the health of workers emerged in the publication in 1700 of a book entitles *Treatise on the Diseases of Artisans* by Bernardino Ramazzini, a professor of medicine at the University of Padua. Ramazzini made a study of occupational diseases and advocated safety measures for workers. His work led eventually to the introduction of factory legislation and compensation for workers.

In Britain, the transition in working methods brought about by the Industrial Revolution resulted in horrendous conditions of work in factories and mines. Women and young children were employed in heavy industry, in mining, and in factories where unguarded machinery posed a daily hazard. Darkness, noise, long hours, polluted atmospheres, and hard taskmasters made the worker's lot far from happy (or safe!). Under the common law, an employer – even in those days – was obliged to provide a reasonably safe place of work, safe machinery and tools in a good state of repair, and provide safety instructions and notice of dangerous conditions. Proving an employer's violation of these common law requirements was by no means easy. When injured at work, in order to sue an employer successfully, the worker had to prove that his employer was personally responsible. The injured worker's inability to obtain any redress or compensation from an employer was almost impossible due to the defences of **contributory negligence** and **common employment. Contributory negligence**, at that time, was the legal presumption that a worker knew and had assumed the risks of the work involved. The doctrine of common employment was a common law expedient which allowed an employer to escape liability for injuries to one of his workers caused by the negligence of a fellow employee. Of course, in those days factory owners were often local magistrates or judges so there was little chance of successful litigation. It is an interesting fact that the common employment situation was still law until repealed by the *Law Reform (Personal Injuries) Act 1948.*

Health and safety legislation began in 1802. The object of the first Act of Parliament, the *Health and Morals of Apprentices Act,* was to stop the employment of children in some 3000 textile mills. It also tried to give them a basic education and minimise the chances of injury.

Some thirty years later, a Royal Commission recommended that inspectors were appointed and, together with the help of the medical profession, they were enlisted to advise and examine employees. The four inspectors were given the right to enter factories and to question workers. Another of their duties was to inspect the free education system. This was the origin of today's health and safety inspectors who since 1974, have operated under the Health and Safety Executive.

From 1833 to 1963 there was a great deal of legislation on safety at work. In 1842 it became illegal to employ children under the age of 15 in certain areas of coal-mines. In 1844 women and young children were prohibited from cleaning any transmission machinery. In 1880 with the introduction of the *Employers' Liability Act*, some employers slowly began to accept liability for employee safety. However, in many cases if you wanted to work you had to forego your rights under the Act and assume any inherent risks of the job in hand.

In 1891, 22 Codes in relation to machinery were passed. Most of this legislation came about in response to particular problems in certain industries. If there was an accident in a mine, legislation was enacted to try to prevent it happening again. However, this reactive and piecemeal approach failed miserably to address the wider safety issues in industry at large. The following synopsis shows the evolution and development of health and safety legislation from 1800 – 1974:

1802 **The Health and Morals of Apprentices Act.** Designed to protect child labour. Working hours limited to 12 hours daily, limited standards were set on heating, lighting and ventilation. The sleeping quarters of boys and girls had to be separated. Some mandatory education of a religious nature was also provided. Enforcement of the Act rested with local magistrates (who were often the guilty parties) and the new law was not effective.

1833 **A Royal Commission** was appointed to investigate child labour in factories. The 1802 Act (together with additions in 1819, 1825 and 1831) was still ineffective. The appointment of the first four "Factory Inspectors" and the institution of a 10 hour day for young persons between the age of 13 and 18 years was the direct result of the Commission's recommendations. However, there was still no such legislation for people outside of the textile industry.

1837 The first common law claim for damages by a worker, for injuries sustained at work, failed.

1842 **The Mines Regulation Act**. Woman and children were no longer permitted to work underground.

1843 The first Mines Inspector was appointed under the 1842 Act.

1860 Factory legislation was gradually extended beyond textiles, but it was still largely ineffective.

1871 The Third Trades Union Congress supported Plimsoll in the debate on "coffin ships".

1880 Employer's Liability Act.

1901 The Factories and Workshops Consolidation Act. A fairly comprehensive code for health and safety, much of which was in force until 1937.

1916 The first **Factories Act** to lay down regulations on the guarding of "prime movers". Also introduce the first regulations on the use of electricity.

1937 Improved **Factories Act.**

1946 National Insurance (Industrial Injuries) Act.

1946 Improved **Factories Act.**

1954 Mines and Quarries Act. Workers given the right to participate in health and safety inspections, etc.

1957 Occupiers' Liability Act. Placed a duty of care on occupiers of premises to safeguard any person lawfully on premises from any dangers emanating from the premises or the activities carried on therein. (Revised 1984)

1961 Factories Act. This amended and consolidated previous Acts..

1963 The first act affecting offices. **The Offices, Shops and Railway Premises Act** gave protection to a further 8 million employees.

1971 Fire Precautions Act. Regulations required that certain premises should have a fired certificate.

1974 The Health and Safety at Work etc Act. Resulting from the Robens' Report, the Act was brought in to cover some 6-8 million additional workers in hospitals and schools etc, plug the loopholes in existing laws, attempt to bring under one umbrella all existing safety laws, and to extend the powers of enforcement agencies as well as affording a measure of protection to members of the public.

1974 Control of Substances Hazardous to Health Regulations. Placed on employers obligations concerning the handling and use of chemicals in the workplace. Early use of risk assessment procedure. (Revised 2004)

1977 Safety Representatives and Safety Committees Regulations. Gave trade unions the right to appoint employee safety representatives and safety committees.

1980 Control of Lead at work Regulations. Required the use of risk assessment to protect health of workers exposed to lead and lead compounds by inhalation, ingestion, or absorption. Placed emphasis on plant control measures rather than individual protection. Introduced medicals and health surveillance for workers. (Revised 2002)

1981 Health and Safety (First Aid) Regulations. Outlined the first aid facilities to be provided in the workplace based on employee numbers and a risk assessment of the degree of risk at the worksite.

1984 Occupiers' Liability Act. Amended the 1957 Act, extending the duty of care to trespassers on premises and particularly children.

1985 Ionising Radiation Regulations. Placed duties on employers to protect employees and other persons from the effects of radiation arising from work with radioactive substances.

1987 Control of Asbestos at Work. Applied to all workplaces where there is a potential risk of exposure to asbestos from its manufacture, use or handling. Introduced absolute prohibition on work with asbestos until proper risk assessment has been made; notification to enforcing authority; proper training; use of PPE and RPE; introduction of asbestos zones; air monitoring; health surveillance and recording; and arrangements for storage, handling and washing facilities. (Revised 2006)

1989 Electricity at Work Regulations. Applied to all work associated with electricity requiring that suitable precautions must be taken against the risk of death or injury from electricity.

1989 Control of Noise at Work Regulations. Introduced levels of noise exposure and action levels in the workplace based on noise assessment. Placed a duty on employers to reduce the risk of hearing damage to employers due to exposure to noise. (Revised 2005)

1990 Environmental Protection Act. Introduced to deal with the whole problem of pollution arising from industrial, domestic and community sources by dealing with discharges to atmosphere, water courses, and waste disposal on land.

1992 Saw the introduction of European-generated health and safety legislation – including the following six sets of regulations, collectively known as the **"six pack"**:

i) **Management of Health and Safety at Work Regulations.** Re-emphasised the principles of the HSWA 1974. Introduced, more generally, the concept of risk assessment in the workplace; health surveillance; consultation; provision of information to employees; arrangements for shared occupancy; control of contractors; employee training; use of temporary workers; employment of young persons and pregnant or nursing females.

ii) **Health and Safety (Display Screen Equipment) Regulations.** Introduced arrangements for dealing with display screen equipment including risk assessment and provision of eye/eyesight testing and spectacles at employer's expense.

iii) **Manual Handling Operations Regulations.** Introduced to cope with manual handling injuries at work based on risk assessments for all manual handling

145

operations and provision of training and instruction in lifting and handling techniques.

iv) **Workplace (Health, Safety and Welfare) Regulations.** Covered all workplaces regarding maintenance, cleanliness, atmosphere, temperature, space, lighting, housekeeping, etc.

v) **Personal Protective Equipment at Work Regulations.** Introduced the provision of protective equipment at employers' expense based on risk assessment – bearing in mind the 'last resort' principle.

vi) **Provision and Use of Work Equipment.** Replaced much of the Factory Act. Covered the safe use at work of every item of plant, equipment and tools.

1994 **Chemical (Hazard Information and Packaging for Supply) Regulations (CHIP).** Regulations to ensure the safe packaging and labelling of chemicals for sale or supply to employers and consumers and provision of safety data for users. (Revised 1996, 2002, 2005)

1994 **Construction (Design and Management) Regulations.** Placed responsibility for safety on construction sites on everyone involved – designer, client, principal and sub-contractor.

1995 **Disability Discrimination Act.** Provided that no person should be discriminated against (in employment) because of any disability.

1995 **Reporting of Injuries, Diseases and Dangerous Occurrences Regulations.** Imposed statutory duty on employers to report accidents, injuries, diseases and dangerous occurrences to the enforcing authority within specified time limits.

1996 **Construction (Heath, Safety and Welfare) Regulations.** Applied to construction work the standards set out in the

Workplace Regulations (see above) and placed duties on main contractor, sub-contractor, site controller and workers. Dealt with platforms, ladders, roof-working, excavations, demolition, vehicles, fire etc and the provision of an emergency plan and site inspections.

1996 Health and Safety (Consultation with Employees) Regulations. Provided consultation rights for management-appointed safety representatives.

1996 Carriage of Dangerous Goods (Classification, Packaging and Labelling) and Use of Transportable Pressure Receptacles Regulations. Dealt with the transport of dangerous substances by rail and road to enable emergency services and public to identify hazards emerging from accidents or spillages (see CHIP above).

1996 Health and Safety (Safety Signs and Signals) Regulations. Specified the type, shape and colour of danger warning signs for use in the workplace and provided information concerning hand signals to be used in connection with handling and lifting operations where the operator has restricted view of the work area.

1997 Fire Safety (Workplace) Regulations. Required employers to carry out fire risk assessments; provide adequate means for fire detection; ensure the provision of adequate means of escape and supply suitable and properly maintained fire equipment.

1997 Confined Spaces Regulations. Prohibited work in a confined space if it could be carried out by other means. Work in a confined space could only be carried out in connection with a safe system of work.

2004 Control of Substances Hazardous to Health (Amendment) Regulations. The main amendment affecting small businesses was to replace maximum exposure limit and occupational exposure standard with a single workplace exposure limit.

2005 Working at Height Regulations. Gave minimum safety and health requirements for use of work equipment (ladders, scaffolding, etc.).

2005 Chemical (Hazard Information and Packaging for Supply) (Amendment) Regulations. Amended the definition of "the approved supply list" contained in the main Regulations. The approved supply list is a list of classified dangerous substances which is added to at various times.

2005 Control of Noise At Work Regulations. These completely replaced the 1989 Regulations but retained much of the early requirements in relation to – exposure limits; risk assessment; hearing protection; health surveillance and training.

2006 Control of Asbestos At Work. These completely replaced the 1987 Regulations and 5 other sets of "Asbestos" Regulations relating to licensing and prohibitions. They retain all the requirements of the earlier Regulations whilst consolidating with them the licensing requirements.

Health and safety legislation is constantly changing. The latest information and up-dates can be obtained by accessing the HSE website (www.hse.gov.uk).

26

In Conclusion

There is no doubt that health and safety suffers from a bad press brought about, in the main, by the crass stupidity of the politically correct in all levels of government both local and national.

Health and safety laws arose out of the terrible working conditions of the 18th and 19th centuries and were aimed at eliminating the widespread evils of that era. However, reducing working hours and preventing the employment of children underground and "up chimneys" is a far cry from stopping the modern child from playing competitive school sport and outlawing the playing of "conkers" in the playground -unless suitably attired in goggles and body armour!

Critics of health and safety focus on the "Nanny State" attitude of those who seek to manage everyday activities through legislation and it is difficult to disagree with that criticism. At the same time, the critic mustn't lose sight of the fact that during the last 20 years thousands of lives were lost in the UK due to work-related accidents or public disasters.

A look at the latest health and safety statistics (2006/2007) shows that 2.2 million people were suffering from illness caused by current or past work. In 2005, 2037 people died from mesothelioma (lung disease); 241 workers were killed at work; 141,350 injuries were reported under RIDDOR; and 36 million working days were lost due to sickness. Also during the period, there were 1141 prosecutions for health and safety offences and the average fine imposed was £8,723.

It is easy to imagine the new employer and small business entrepreneur disappearing under a barrage of red tape. The key is to adopt a balanced perspective and approach health and safety as just another aspect of business which can be managed quite easily by the application of a large dose of common sense.

The appendices include information on various Acts of Parliament, statutory instruments, codes of practice and a list of useful abbreviations and acronyms. Legislation is notoriously difficult to interpret accurately. Codes of practice (and you will very likely find one appropriate to your business) whilst not exactly law, are much easier to understand. Rest assured, if you are managing your business according to the advice given in a code of practice you are on safe ground.

You can also obtain up to date information from HSE, HMSO and SME websites.

Appendix 1
Health and Safety Legislation

Chemicals (Hazard Information and Packaging for Supply) Regulations 2002/2005

Construction (Design and Management) Regulations 2007

Construction (Health, Safety and Welfare) Regulations 1996

Control of Asbestos Regulations 2006

Control of Lead at Work Regulations 2002

Control of Major Accident Hazards Regulations 1999

Control of Noise at Work Regulations 2005

Control of Substances Hazardous to Health Regulations 2002/2004

Corporate Manslaughter and Corporate Homicide Act 2007

Disability Discrimination Act 2005

Electricity at Work Regulations 1989/2006

Employer's Liability (Compulsory Insurance) Act 1969

Employers' Liability (Compulsory Insurance) Regulations 1998

Employment Protection Act 1975

Environmental Protection Act 1990

Factories Act 1961

Fire Precautions Act 1971

Fire Safety (Workplace) Regulations 1997

Food Safety Act 1990

Hazardous Waste (England and Wales) Regulations 2005

Health and Safety (Display Screen Equipment) Regulations 1992

Health and Safety (First-Aid) Regulations 1981/2002

Health and Safety (Signs and Signals) Regulations 1996

Health and Safety at Work etc Act 1974

Information and Consultation of Employees) Regulations 2004/2006

Ionising Radiations Regulations 1985 and 1999

Management of Health and Safety at Work Regulations 1992 and 1999

Manual Handling Operations Regulations 1992

Occupier's Liability Act 1957 and 1984

Offices, Shops and Railway Premises Act 1963

Personal Protective Equipment at Work Regulations 1992

Provision and Use of Work Equipment Regulations 1992 and 1998

Regulatory Reform (Fire Safety) Order 2005

Reporting of Injuries, Diseases and Dangerous Occurrences Regulations 1995

Safety Representatives and Safety Committees Regulations 1977

Social Security (Industrial Diseases) (Prescribed Diseases) Regulations 1980

Special Waste Regulations 1996

Supply of Machinery (Safety) Regulations 1992/2005

Work at Height Regulations 2005

Working Time Regulations 1998/2007

Workplace (Health, Safety and Welfare) Regulations 1992

NB. The above list is not exhaustive and legislation is constantly being revised, consolidated and revoked. The reader should consult HMSO for details of current legislation.

Appendix 2
Approved Codes of Practice

COP Series

Control of lead at work – approved code of practice.	COP 2
Standards of training in safe gas installation – approved code of practice 1987.	COP 20
Safety in Docks: Docks Regulations 1988 – approved code of practice.	COP 25
Safety of exit from mines underground workings – approved code of practice 1988.	COP 28
First aid on offshore installations and pipeline works – approved code of practice with regulations and guidance 1990.	COP 32
Safety of pressure systems: Pressure Systems and Transportable Gas Containers Regulations 1989 – approved code of practice 1989.	COP 37

L Series

A Guide to the Health and Safety at Work etc Act 1974: guidance on the Act.	L 1
Management of health and safety at work. Management of Health and Safety at Work Regulations 1999 – approved code of practice 2000.	L 21
Workplace health, safety and welfare. Workplace (Health, safety and Welfare) Regulations 1992 – approved code of practice and guidance 1992.	L 24
The control of asbestos at work: Control of Asbestos at Work Regulations 1987 – approved code of practice 1999.	L 27
First aid at work: The Health and safety (First aid) Regulations 1981 – approved code of practice and guidance 1997.	L 74
Work with ionising radiation: Ionising Radiations Regulations 1999 – approved code of practice.	L 121

NB The above lists are not exhaustive and the reader should consult HSE Books for up-to-date and current information on publications.

Appendix 3
Additional Information Concerning COSHH, CHIP and EH40

1. COSHH

The following are selected definitions from Regulation 2 of the Control of Substances Hazardous to Health Regulations 2002:

biological agent   a micro-organism, cell culture, or human endoparasite, whether or not genetically modified, which may cause infection, allergy, toxicity or otherwise create a hazard to human health.

carcinogen   (a) a substance or preparation which if classified in accordance with the classification provided for by regulation 4 of the CHIP Regulations would be in the category of danger, carcinogenic (category 1) or carcinogenic (category 2) whether or not the substance or preparation would be required to be classified under those Regulations; or

(b) a substance or preparation –
   (1) listed in Schedule 1, or
   (2) arising froma process specified in Schedule 1 which is a substance hazardous to health.

the CHIP Regulations the Chemicals (Hazard Information and Packaging for Supply) Regulations 2002.

control measure	a measure taken to reduce exposure to a substance hazardous to health including the provision of systems of work and supervision, the cleaning of workplaces, premises, plant and equipment, the provision and use of engineering controls and personal protective equipment).
hazard	in relation to a substance, the intrinsic property of that substance which has the potential to cause harm to the health of a person, and "hazardous" shall be construed accordingly.
health surveillance	assessment of the state of health of an employee, as related to exposure to substances hazardous to health, and includes biological monitoring.
inhalable dust	airborne material which is capable of entering the nose and mouth during breathing, as defined by BS EN 481 1993.
micro-organism	a microbiological entity, cellular or non-cellular, which is capable of replicating or of transferring genetic material.
personal protective equipment	all equipment (including clothing) which is intended to be worn or held by a person at work and which protects that person against one or more risks to his health, and any additional accessory designed to meet that objective.
preparation	a mixture or solution of two or more substances.
respirable dust	airborne material which is capable of penetrating to the gas exchange region of the

lung, as defined by BS EN 481 1993.

risk

in relation to the exposure of an employee to a substance hazardous to health, the likelihood that the potential for harm to the health of a person will be attained under the conditions of use and exposure and also the extent of that harm.

substance

a natural or artificial substance whether in solid or liquid form or in the form of a gas or vapour (including micro-organisms).

substance hazardous to health

a substance (including a preparation) –
(a) which is listed in Part 1 of the approved supply list as dangerous for supply within the meaning of the CHIP Regulations and for which an indication of danger specified for the substance is very toxic, toxic, harmful, corrosive or irritant;
(b) for which the Health and Safety Commission has approved a workplace exposure limit;
(c) which is a biological agent;
(d) which is dust of any kind, except dust which is a substance within paragraph (a) or (b) above, when present at a concentration in air equal to or greater than –
(i) 10mg/m3 as a time-weighted average over an 8 hour period, of inhalable dust, or
(ii) 4mg/m3 as a time-weighted average over an 8 hour period, of respirable dust;
(e) which, not being a substance falling within sub-paragraphs (a) to (d), because of its chemical or toxicological properties and the way it is used or is present at the workplace creates a risk to health.

workplace	any premises or part of premises used for or in connection with work, and includes – (a) any place within the premises to which an employee has access while at work; and (b) any room, lobby, corridor, staircase, road or other place – (i) used as a means of access to or egress from that place of work, or (ii) where facilities are provided for use in connection with that place of work, other than a public road.

2. CHIP

The following is reproduced from Schedules 1 and 2 of the Chemicals (Hazard Information and Packaging for Supply) Regulations 2002.

Classification of dangerous substances and dangerous preparations

Category of danger	Property
Explosive	Solid, liquid pasty or gelatinous substances and preparations which may react exothermically without atmospheric oxygen thereby quickly evolving gases, and which under defined test conditions detonate, quickly deflagrate or upon heating explode when partially confined.
Oxidising	Substances and preparations which give rise to a highly exothermic reaction in contact with other substances, particularly flammable substances.
Extremely flammable	Substances and preparations having an extremely low flash point and a low boiling point and gaseous substances and preparations which are flammable in contact with air at ambient temperature and pressure.

Highly flammable	The following substances and preparations, namely (a) substances and preparations which may become hot and finally catch fire in contact with air at ambient temperature without any application of energy (b) solid substances and preparations which may readily catch fire after brief contact with a source of ignition and which continue to burn or to be consumed after removal of the source of ignition (c) liquid substances and preparations having a very low flash point, or (d) substances and preparations which, in contact with water or damp air, evolve extremely flammable gases in dangerous quantities.
Flammable	Liquid substances and preparations having a low flash point.
Very toxic	Substances and preparations which in very low quantities cause death or acute or chronic damage to health when inhaled, swallowed or absorbed via the skin.
Toxic	Substances and preparations which in low quantities cause death or acute or chronic damage to health when inhaled, swallowed or absorbed via the skin.
Harmful	Substances and preparations which may cause death or acute or chronic damage to health when inhaled, swallowed or absorbed via the skin.
Corrosive	Substances and preparations which may, on contact with living tissues, destroy them.
Irritant	Non-corrosive substances and preparations which, through immediate, prolonged or repeated contact with the skin or mucous membrane, may cause inflammation.

3. EH40

Copies of the publications mentioned above are obtainable as follows:

a) a list of the workplace exposure limits which the Health and Safety Commission has approved is available in the publication "EH40, Occupational Exposure Limits" obtainable from HSE Books, PO Box1999, Sudbury Suffolk CO10 2WA; and

b) British Standard BS EN 481 1993, referred to in Regulation 2(i) of the Chemicals (Hazard Information and Packaging for Supply) Regulations, relating to workplace atmospheres, is obtainable from the British Standards Institute, BSI House, 389 Chiswick High Road, London W4 4AL.

Appendix 4
Health and Safety Guidance Notes

Chemical Safety Series

Storage and use of sodium chlorate and other similar strong oxidants 1998	CS 3
Storage and use of LPG on metered estates 1987	CS 11
The cleaning and gas freeing of tanks containing flammable residues 1985	CS 15
Storage and handling of ammonium nitrate 1986	CS 18
Storage and handling of organic peroxides 1991	CS 21
Fumigation 1996	CS 22

Environmental Hygiene Series

Cadmium: health and safety precautions 1995	EH 1
Chromium and its inorganic compounds: health and safety precautions 1998	EH 2
Asbestos: exposure limits and measurement of airborne dust concentrations 1995	EH 10
Beryllium: health and safety precautions 1995	EH 13
Isocyanates: health hazards and precautionary measures 1999	EH 16
Mercury and its inorganic divalent compounds 1996	EH 17

General Series

Safety in pressure testing 1998	GS 4
Avoidance of danger from overhead electrical lines 1997	GS 6
Safe erection of structures: Part 2. Site management and procedures 1985	GS 28/2
Health and safety in shoe repair premises 1984	GS 32
Electrical test equipment for use by electricians 1995	GS 38
Pre-stressed concrete 1991	GS 49

Medical Series

Colour vision 1987	MS 7
Mercury: medical guidance notes 1996	MS 12
Asbestos: medical guidance notes 1996	MS 13
Biological monitoring of workers exposed to organo-phosphorous pesticides 2000	MS 17
Health surveillance of occupational skin disease 1998	MS 24
Medical aspects of occupational asthma 1998	MS 25

Plant and Machinery Series

High temperature textile dyeing machines 1997	PM 4
Automatically controlled steam and hot water boilers 1989	PM 5
Safety in the use of pallets	PM 15
Eyebolts 1978	PM 16
Pneumatic nailing and stapling guns 1979	PM 17
Safety at rack and pinion hoists 1981	PM 24

Legal Series (L)

See Approved Codes of Practice – Appendix 2

N.B. The above lists are not exhaustive and the reader should consult HSE Books for details of up-to-date and current publications.

Appendix 5
Health and Safety Guidance Booklets

Safety advice for bulk chlorine installations	HSG 28
Pie and tart machines	HSG 31
Anthrax	HSG 36
An introduction to local exhaust ventilation	HSG 37
Lighting at work	HSG 38
Compressed air safety	HSG 39
Safe handling of chlorine from drums and cylinders	HSG 40
Safety in the use of metal cutting guillotines and shears	HSG 42
Industrial robot safety	HSG 43
Reducing error and influencing behaviour	HSG 48
Storage of flammable liquids in containers	HSG 51
The maintenance, examination and testing of local exhaust ventilation	HSG 54
Seating at work	HSG 57
Health surveillance at work	HSG 61
Health and safety in tyre and exhaust fitting premises	HSG 62
Successful health and safety management	HSG 65
Protection of workers and the general public during the development of contaminated land	HSG 66
Health and safety in motor vehicle repair	HSG 67
Chemical warehousing	HSG 71

Control of respirable silica dust in heavy clay and refractory processes	HSG 72
Control of respirable crystalline silica in quarries	HSG 73
Health and safety in retail and wholesale warehouses	HSG 76
Dangerous goods in cargo transport units	HSG 78
Health and safety in golf course management and maintenance	HSG 79
Electricity at work	HSG 85
Safety in the remote diagnosis of manufacturing plant and equipment	HSG 87
Hand-arm vibration	HSG 88
Safeguarding agricultural machinery	HSG 89
The law on VDUs; An easy guide	HSG 90
Safe use and storage of cellular plastics	HSG 92
The assessment of pressure vessels operating at low temperature	HSG 93
Safety in the design and use of gamma and electron irradiation facilities	HSG 94
The radiation safety of lasers used for display purposes	HSG 95
The costs of accidents at work	HSG 96
A step by step guide to COSHH assessment	HSG 97
Prevention of violence to staff in banks and building societies	HSG 100
The costs to Britain of workplace accidents and work-related ill health in 1995/96	HSG 101
Safe handling of combustible dusts	HSG 103
Control of noise in quarries	HSG 109
Maintaining portable and transportable electrical equipment	HSG 107
Seven steps to successful substitution of hazardous substances	HSG 110
Seven steps to successful substitution of hazardous substances (in Welsh)	HSG 110W

Lift trucks in potentially flammable atmospheres	HSG 113
Conditions for the authorisation of explosives in Great Britain	HSG 114
Manual Handling	HSG 115
Making sense of NONS	HSG 117
Electrical safety in arc welding	HSG 118
Manual handling in drinks delivery	HSG 119
A pain in your workplace?	HSG 121
New and expectant mothers at work	HSG 122
Working together on firework displays	HSG 123
Giving your own firework display	HSG 124
A brief guide on COSHH for the offshore oil and gas Industry	HSG 125
Energetic and spontaneously combustible substances	HSG 131
How to deal with sick building syndrome	HSG 132
Preventing violence to retail staff	HSG 133
Storage and handling of industrial nitrocellulose	HSG 135
Workplace transport safety	HSG 136
Health risk management	HSG 137
Sound solutions	HSG 138
Safe use of compressed gases in welding, flame cutting and allied processes	HSG 139
Safe use and handling of flammable liquids	HSG 140
Electrical safety on construction sites	HSG 141
Dealing with offshore emergencies	HSG 142
Dispensing petrol	HSG 146
Health and safety in construction	HSG 150
Protecting the public	HSG 151
Railway safety principles and Guidelines. Part 1	HSG 153/1

Railway safety principles and guidance. Part 2 Section A	HSG 153/2
Railway safety principles and Guidelines. Part 2. Section C	HSG 153/4
Railway safety principles and guidance . Part 2 Section D	HSG 153/5
Railway safety principles and guidance. Part 2 Section E	HSG 153/6
Railway safety principles and guidance. Part 2. Section F	HSG 153/7
Railway safety principles and guidance. Part 2. Section G	HSG 153/8
Managing crowds safely	HSG 154
Slips and trips	HSG 155
Slips and trips	HSG 156
Flame arresters	HSG 158
The carriage of dangerous goods explained. Part 1	HSG 160
The carriage of dangerous goods explained. Part 2	HSG 161
The carriage of dangerous goods explained	HSG 162
The carriage of dangerous goods explained Part 3	HSG 163
The carriage of dangerous goods explained Part 5	HSG 164
Formula for health and safety	HSG 166
Biological monitoring in the workplace	HSG 167
Fire safety in construction work	HSG 168
Camera operations on location	HSG 169
Vibration solutions	HSG 170
Well handled	HSG 171
Health and safety in sawmilling	HSG 172
Monitoring strategies for toxic substances	HSG 173
Fairgrounds and amusement parks. Guidance on safe practice	HSG 175
Storage of flammable liquids in tanks	HSG 176

The spraying of flammable liquids	HSG 178
Application of electro-sensitive protective equipment using light curtains and light beam devices to machinery	HSG 180
Assessment principles for offshore safety cases	HSG 181
Sound solutions offshore	HSG 182
5 steps to risk assessment	HSG 183
Guidance on the handling, storage and transport of airbags and seatbelt pretensioners	HSG 184
The bulk transfer of dangerous liquids and gases between ship and shore	HSG 186

N.B. The above lists are not exhaustive and the reader should consult HSE Books for details of up-to-date and current publications.

Appendix 6
Abbreviations and Acronyms

The following is a list of abbreviations and acronyms which are common usage in health and safety matters

ACAS	Advisory Conciliation and Arbitration Service
ACOP	Approved Code of Practice
ACTS	Advisory Committee on Toxic Substances
ADR	European Agreement on International Carriage of Dangerous Goods by Road
AIDS	Acquired Immunodeficiency Syndrome
APC	Air Pollution Control
BATNEEC	Best available techniques not entailing excessive cost
BI 510	Accident Book (Approved)
BLEVE	Boiling liquid expanding vapour explosion
BNFL	British Nuclear Fuels
BPEO	Best practical environmental option
BS	British Standard
BSC	British Safety Council
BSI	British Standards Institute
CATNAP	Cheapest available technique not attracting prosecution
CDM	Construction Design and Management Regulations
CE	Commission Europeene (acute accent)
CEN	Committee Europeen de Normalisation

CENELEC	Committee Europeen de Normalisation Electrotechnique
CFC	Chlorofluorocarbon
CHIP	Chemicals (Hazard Information and Packaging for Supply) Regulations 2002
CITB	Construction Industry Training Board
COMAH	Control of Major Accident Hazards Regulations
COSHH	Control of Substances Hazardous to Health Regulations 2002
dB	Decibel
dB(A)	'A' weighted decibel – Occupational noise unit of measurement
DEFRA	Department for Environment, Food and Rural Affairs
DTI	Department of Trade and Industry
DDA	Disability Discrimination Act 1995
DDPA	Disability Discrimination (Premises) Act 2006
DSE	Display Screen Equipment Regulations
EAT	Employment Appeals Tribunal
EC	European Commission
EEC	European Economic Community
EHO	Environmental Health Officer
EINECS	European Inventory of Existing Chemical Substances
ELCD	Earth leakage circuit breaker
ELINCS	European List of Notified Chemical Substances
EMAS	Employment Medical Advisory Service
EN	European Normalisation prefix
EU	European Union
FLT	Fork lift Truck
FMEA	Failure Mode and Effect Analysis

HACCP	Hazard analysis and critical control points
HAZCHEM	Hazardous chemicals code
HAZOP	Hazard and operability study
HSC	Health and Safety Commission
HMIP	Her Majesty's Inspectorate of Pollution
HMSO	Her Majesty's Stationery Office
HFL	Highly flammable liquid
HIV	Human Immunodeficiency Virus
HSE	Health and Safety Executive
HSWA	Health and Safety at Work etc Act 1974
Hz	Hertz
IEE	Institute of Electrical Engineers
IOSH	Institute of Occupational Safety and Health
IPC	Integrated Pollution Control
ISO	International Standards Organisation
IUPAC	International Union of Pure and Applied Chemistry
LASER	Light amplification by stimulated emission of radiation
LC50	Lethal concentration
LD50	Lethal dose
LEV	Local exhaust ventilation
LOLER	Lifting Operations and Lifting Equipment Regulations
LPG	Liquid Petroleum Gas
LEV	Local exhaust ventilation
MHSW	Management of Health and Safety at Work Regulations1999
MHOR	Manual Handling Operations Regulations1992
MEL	Maximum exposure limit
MIIRSM	Member of the International Institute of Risk and Safety Management

NEBOSH	National Examination Board in Occupational Safety and Health
NICEIC	National Inspection Council for Electrical Installation Contracting
NRPB	National Radiation Protection Board
OES	Occupational Exposure Standard
pH	Measure of acidity/alkalinity
PPE	Personal Protective Equipment
PUWER	Provision and Use of Work Equipment Regulations 1998
PVC	Polyvinyl chloride
RCD	Residual current device
REM	Roentgen Equivalent Man
RIDDOR	Reporting of Injuries Diseases and Dangerous Occurrence Regulations 1995
RoSPA	Royal Society for the Prevention of Accidents
RPE	Respiratory Protective Equipment
RSI	Repetitive Strain Injury
STEL	Short Term Exposure Limit
SWL	Safe working load
TLV	Threshold limit value
TREMCARD	Transport emergency card
TWA	Time weighted average
WRULD	Work-related upper limb disorder
WTR	Working Time Regulations

Appendix 7
Special Wastes

Special wastes are defined in Schedule 2 of the Special Waste Regulations 1996 as follows:

Waste code	Description
02	WASTE FROM AGRICULTURAL, HORTICULTURAL, HUNTING, FISHING AND AQUACULTURE PRIMARY PRODUCTION, FOOD PREPARATION AND PROCESSING
0201	PRIMARY PRODUCTION WASTE
020105	agrochemical wastes
03	WASTES FROM WOOD PROCESSING AND THE PRODUCTION OF PAPER, CARDBOARD, PULP, PANELS AND FURNITURE
0302	WOOD PRESERVATION WASTE
030201	non-halogenated organic wood preservatives
030202	organochlorinated wood preservatives
030203	organometallic wood preservatives
030204	inorganic wood preservatives
04	WASTES FROM THE LEATHER AND TEXTILE INDUSTRIES
0401	WASTES FROM THE LEATHER INDUSTRY
040103	degreasing wastes containing solvents without a liquid phase
0402	WASTES FROM TEXTILE INDUSTRY
040211	halogenated wastes from dressing and finishing
05	WASTES FROM PETROLEUM REFINING, NATURAL GAS PURIFICATION AND PYROLYTIC TREATMENT OF COAL
0501	OILY SLUDGES AND SOLID WASTES
050103	tank bottom sludges
050104	acid alkyl sludges
050105	oil spills
050107	acid tars
050108	other tars
0504	SPENT FILTER CLAYS
050401	spent filter clays
0506	WASTE FROM THE PYROLYTIC TREATMENT OF COAL
050601	acid tars
050603	other tars
0507	WASTE FROM NATURAL GAS PURIFICATION
050701	sludges containing mercury
0508	WASTES FROM OIL REGENERATION
050801	spent filter clays
050802	acid tars
050803	other tars

050804	aqueous liquid waste from oil regeneration
06	WASTES FROM INORGANIC CHEMICAL PROCESSES
0601	WASTE ACIDIC SOLUTIONS
060101	sulphuric acid and sulphurous acid
060102	hydrochloric acid
060103	hydrofluoric acid
060104	phosphoric and phosphorous acid
060105	nitric acid and nitrous acid
060199	waste not otherwise specified
0602	ALKALINE SOLUTIONS
060201	calcium hydroxide
060202	soda
060203	ammonia
060299	wastes not otherwise specified
0603	WASTE SALTS AND THEIR SOLUTIONS
060311	salts and solutions containing cyanides
0604	METAL-CONTAINING WASTES
060402	metallic salts (except 0603)
060403	wastes containing arsenic
060404	wastes containing mercury
060405	wastes containing heavy metals
0607	WASTES FROM HALOGEN CHEMICAL PROCESSES
060701	wastes containing asbestos from electrolysis
060702	activated carbon from chlorine production
0613	WASTES FROM OTHER INORGANIC CHEMICAL PROCESSES
061301	inorganic pesticides, biocides and wood preserving agents
061302	spent activated carbon (except 060702)
07	WASTES FROM ORGANIC CHEMICAL PROCESSES
0701	WASTE FROM THE MANUFACTURE, FORMULATION, SUPPLY AND USE (MFSU) OF BASIC ORGANIC CHEMICALS
070101	aqueous washing liquids and mother liquors
070103	organic halogenated solvents, washing liquids and mother liquors
070104	other organic solvents, washing liquids and mother liquors
070107	halogenated still bottoms and reaction residues
070108	other still bottoms and reaction residues
070109	halogenated filter cakes, spent absorbents
070110	other filter cakes, spent absorbents
0702	WASTE FROM THE MFSU OF PLASTICS, SYNTHETIC RUBBER AND MAN-MADE FIBRES
070201	aqueous washing liquids and mother liquors
070203	organic halogenated solvents, washing liquids and mother liquors
070204	other organic solvents, washing liquids and mother liquors
070207	halogenated still bottoms and reaction residues
070208	other still bottoms and reaction residues
070209	halogenated filter cakes, spent absorbents
070210	other filter cakes, spent absorbents
0703	WASTE FROM THE MFSU FOR ORGANIC DYES AND PIGMENTS (EXCLUDING 0611)
070301	aqueous washing liquids and mother liquors
070303	organic halogenated solvents, washing liquids and mother liquors
070304	other organic solvents, washing liquids and mother liquors
070307	halogenated still bottoms and reaction residues
070308	other still bottoms and reaction residues
070309	halogenated filter cakes, spent absorbents
070310	other filter cakes, spent absorbents
0704	WASTE FROM THE MFSU FOR ORGANIC PESTICIDES (EXCEPT 020105)
070401	aqueous washing liquids and mother liquors

173

070403	organic halogenated solvents, washing liquids and mother liquors
070404	other organic solvents, washing liquids and mother liquors
070407	halogenated still bottoms and reaction residues
070408	other still bottoms and reaction residues
070409	halogenated filter cakes, spent absorbents
070410	other filter cakes, spent absorbents
0705	WASTE FROM THE MFSU OF PHARMACEUTICALS
070501	aqueous washing liquids and mother liquors
070503	organic halogenated solvents, washing liquids and mother liquors
070504	other organic solvents, washing liquids and mother liquors
070507	halogenated still bottoms and reaction residues
070508	other still bottoms and reaction residues
070509	halogenated filter cakes, spent absorbents
070510	other filter cakes, spent absorbents
0706	WASTE FROM THE MFSU OF FATS, GREASE, SOAPS, DETERGENTS, DISINFECTANTS AND COSMETICS
070601	aqueous washing liquids and mother liquors
070603	organic halogenated solvents, washing liquids and mother liquors
070604	other organic solvents, washing liquids and mother liquors
070607	halogenated still bottoms and reaction residues
070608	other still bottoms and reaction residues
070609	halogenated filter cakes, spent absorbents
070610	other filter cakes, spent absorbents
0707	WASTE FROM THE MFSU OF FINE CHEMICALS AND CHEMICAL PRODUCTS NOT OTHERWISE SPECIFIED
070701	aqueous washing liquids and mother liquors
070703	organic halogenated solvents, washing liquids and mother liquors
070704	other organic solvents, washing liquids and mother liquors
070707	halogenated still bottoms and reaction residues
070708	other still bottoms and reaction residues
070709	halogenated filter cakes, spent absorbents
070710	other filter cakes, spent absorbents
08	WASTES FROM THE MANUFACTURE, FORMULATION, SUPPLY AND USE (MFSU) OF COATINGS (PAINTS, VARNISHES AND VITREOUS ENAMELS), ADHESIVE, SEALANTS AND PRINTING INKS
0801	WASTES FROM MFSU OF PAINT AND VARNISH
080101	waste paints and varnish containing halogenated solvents
080102	waste paints and varnish free of halogenated solvents
080106	sludges from paint or varnish removal containing halogenated solvents
080107	sludges from paint or varnish removal free of halogenated solvents
0803	WASTES FROM MFSU OF PRINTING INKS
080301	waste ink containing halogenated solvents
080302	waste ink free of halogenated solvents
080305	ink sludges containing halogenated solvents
080306	ink sludges free of halogenated solvents
0804	WASTES FROM MFSU OF ADHESIVE AND SEALANTS (INCLUDING WATER-PROOFING PRODUCTS)
080401	waste adhesives and sealants containing halogenated solvents
080402	waste adhesives and sealants free of halogenated solvents
080405	adhesives and sealants sludges containing halogenated solvents
080406	adhesives and sealants sludges free of halogenated solvents
09	WASTES FROM THE PHOTOGRAPHIC INDUSTRY
0901	WASTES FROM PHOTOGRAPHIC INDUSTRY
090101	water based developer and activator solutions
090102	water based offset plate developer solutions
090103	solvent based developer solutions
090104	fixer solutions

Appendix 7

090105	bleach solutions and bleach fixer solutions
090106	waste containing silver from on-site treatment of photographic waste
10	INORGANIC WASTES FROM THERMAL PROCESSES
1001	WASTES FROM POWER STATION AND OTHER COMBUSTION PLANTS (EXCEPT 1900)
100104	oil fly ash
100109	sulphuric acid
1003	WASTES FROM ALUMINIUM THERMAL METALLURGY
100301	tars and other carbon-containing wastes from anode manufacture
100303	skimmings
100304	primary smelting slags/white drosses
100307	spent pot lining
100308	salt slags from secondary smelting
100309	black drosses from secondary smelting
100310	waste from treatment of salt slags and black drosses treatment
1004	WASTES FROM LEAD THERMAL METALLURGY
100401	slags (1st and 2nd smelting)
100402	dross and skimmings (1st and 2nd smelting)
100403	calcium arsenate
100404	flue gas dust
100405	other particulates and dust
100406	solid waste from gas treatment
100407	sludges from gas treatment
1005	WASTES FROM ZINC THERMAL METALLURGY
100501	slags (1st and 2nd smelting)
100502	dross and skimmings (1st and 2nd smelting)
100503	flue gas dust
100505	solid waste from gas treatment
100506	sludges from gas treatment
1006	WASTES FROM COPPER THERMAL METALLURGY
100603	flue gas dust
100605	waste from electrolytic refining
100606	solid waste from gas treatment
100607	sludges from gas treatment
11	INORGANIC WASTE WITH METALS FROM METAL TREATMENT AND THE COATING OF METALS; NON-FERROUS HYDRO-METALLURGY
1101	LIQUID WASTES AND SLUDGES FROM METAL TREATMENT AND COATING OF METALS (e.g. GALVANIC PROCESSES, ZINC COATING PROCESSES, PICKLING PROCESSES, ETCHING, PHOSPHATIZING, ALKALINE DE-GREASING)
110101	cyanidic (alkaline) wastes containing heavy metals other than chromium
110102	cyanidic (alkaline) wastes which do not contain heavy metals
110103	cyanide-free wastes containing chromium
110105	acidic pickling solutions
110106	acids not otherwise specified
110107	alkalis not otherwise specified
110108	phosphatizing sludges
1102	WASTES AND SLUDGES FROM NON-FERROUS HYDROMETALLURGICAL PROCESSES
110202	sludges from zinc hydrometallurgy (including jarosite, goethite)
1103	SLUDGES AND SOLIDS FROM TEMPERING PROCESSES
110301	wastes containing cyanide
110302	other wastes
12	WASTES FROM SHAPING AND SURFACE TREATMENT OF METALS AND PLASTICS
1201	WASTES FROM SHAPING (INCLUDING FORGING, WELDING, PRESSING, DRAWING, TURNING, CUTTING AND FILING)

120106	waste machining oils containing halogens (not emulsioned)
120107	waste machining oils free of halogens (not emulsioned)
120108	waste machining emulsions containing halogens
120109	waste machining emulsions free of halogens
120110	synthetic machining oils
120111	machining sludges
120112	spent waxes and fats
1203	WASTES FROM WATER AND STEAM DEGREASING PROCESSES (EXCEPT 1100)
120301	aqueous washing liquids
120302	steam degreasing wastes
13	OIL WASTES (EXCEPT EDIBLE OILS, 0500 AND 1200)
1301	WASTE HYDRAULIC OILS AND BRAKE FLUIDS
130101	hydraulic oils, containing PCBs or PCTs
130102	other chlorinated hydraulic oils (not emulsions)
130103	non-chlorinated hydraulic oils (not emulsions)
130104	chlorinated emulsions
130105	non-chlorinated emulsions
130106	hydraulic oils containing only mineral oil
130107	other hydraulic oils
130108	brake fluids
1302	WASTE ENGINE, GEAR AND LUBRICATING OILS
130201	chlorinated engine, gear and lubricating oils
130202	non-chlorinated engine, gear and lubricating oils
130203	other machine, gear and lubricating oils
1303	WASTE INSULATING AND HEAT TRANSMISSION OILS AND OTHER LIQUIDS
130301	insulating or heat transmission oils and other liquids containing PCBs or PCTs
130302	other chlorinated insulating and heat transmission oils and other liquids
130303	non-chlorinated insulating and heat transmission oils and other liquids
130304	synthetic insulating and heat transmission oils and other liquids
130305	mineral insulating and heat transmission oils
1304	BILGE OILS
130401	bilge oils from inland navigation
130402	bilge oils from jetty sewers
130403	bilge oils from other navigation
1305	OIL/WATER SEPARATOR CONTENTS
130501	oil/water separator solids
130502	oil/water separator sludges
130503	interceptor sludges
130504	desalter sludges or emulsions
130505	other emulsions
1306	OIL WASTE NOT OTHERWISE SPECIFIED
130601	oil waste not otherwise specified
14	WASTES FROM ORGANIC SUBSTANCES EMPLOYED AS SOLVENTS (EXCEPT 0700 AND 0800)
1401	WASTES FROM METAL DEGREASING AND MACHINERY MAINTENANCE
140101	chlorofluorocarbons
140102	other halogenated solvents and solvent mixes
140103	other solvents and solvent mixes
140104	aqueous solvent mixes containing halogens
140105	aqueous solvent mixes free of halogens
140106	sludges or solid wastes containing halogenated solvents
140107	sludges or solid wastes free of halogenated solvents
1402	WASTES FROM TEXTILE CLEANING AND DEGREASING OF NATURAL PRODUCTS
140201	halogenated solvents and solvent mixes
140202	solvent mixes or organic liquids free of halogenated solvents

140203	sludges or solid wastes containing halogenated solvents
140204	sludges or solid wastes containing other solvents
1403	WASTES FROM THE ELECTRONIC INDUSTRY
140301	chlorofluorocarbons
140302	other halogenated solvents
140303	solvents and solvent mixes free of halogenated solvents
140304	sludges or solid wastes containing halogenated solvents
140305	sludges or solid wastes containing other solvents
1404	WASTES FROM COOLANTS, FOAM/AEROSOL PROPELLANTS
140401	chlorofluorocarbons
140402	other halogenated solvents and solvent mixes
140403	other solvents and solvent mixes
140404	sludges or solid wastes containing halogenated solvents
140405	sludges or solid wastes containing other solvents
1405	WASTES FROM SOLVENT AND COOLANT RECOVERY (STILL BOTTOMS)
140501	chlorofluorocarbons
140502	halogenated solvents and solvent mixes
140503	other solvents and solvent mixes
140504	sludges containing halogenated solvents
140505	sludges containing other solvents
16	WASTES NOT OTHERWISE SPECIFIED IN THE CATALOGUE
1602	DISCARDED EQUIPMENT AND SHREDDER RESIDUES
160201	transformers and capacitors containing PCBs or PCTs
1604	WASTE EXPLOSIVES
160401	waste ammunition
160402	fireworks waste
160403	other waste explosives
1606	BATTERIES AND ACCUMULATORS
160601	lead batteries
160602	Ni-Cd batteries
160603	mercury dry cells
160606	electrolyte from batteries and accumulators
1607	WASTE FROM TRANSPORT AND STORAGE TANK CLEANING (EXCEPT 0500 AND 1200)
160701	waste from marine transport tank cleaning, containing chemicals
160702	waste from marine transport tank cleaning, containing oil
160703	waste from railway and road transport tank cleaning, containing oil
160704	waste from railway and road transport tank cleaning, containing chemicals
160705	waste from storage tank cleaning, containing chemicals
160706	waste from storage tank cleaning, containing oil
17	CONSTRUCTION AND DEMOLITION WASTE (INCLUDING ROAD CONSTRUCTION)
1706	INSULATION MATERIALS
170601	insulation materials containing asbestos
18	WASTES FROM HUMAN OR ANIMAL HEALTH CARE AND/OR RELATED RESEARCH (EXCLUDING KITCHEN AND RESTAURANT WASTES WHICH DO NOT ARISE FROM IMMEDIATE HEALTH CARE)
1801	WASTE FROM NATAL CARE, DIAGNOSIS, TREATMENT OR PREVENTION OF DISEASE IN HUMANS
180103	other wastes whose collection and disposal is subject to special requirements in view of the prevention of infection
1802	WASTE FROM RESEARCH, DIAGNOSIS, TREATMENT OR PREVENTION OF DISEASE INVOLVING ANIMALS
180202	other wastes whose collection and disposal is subject to special requirements in view of the prevention of infection
180204	discarded chemicals

177

19	WASTES FROM WASTE TREATMENT FACILITIES, OFF-SITE WASTE WATER TREATMENT PLANTS AND THE WATER INDUSTRY
1901	WASTES FROM INCINERATION OR PYROLYSIS OF MUNICIPAL AND SIMILAR COMMERCIAL, INDUSTRIAL AND INSTITUTIONAL WASTES
190103	fly ash
190104	boiler dust
190105	filter cake from gas treatment
190106	aqueous liquid waste from gas treatment and other aqueous liquid wastes
190107	solid waste from gas treatment
190110	spent activated carbon from flue gas treatment
1902	WASTES FROM SPECIFIC PHYSICO/CHEMICAL TREATMENTS OF INDUSTRIAL WASTES (e.g. DECHROMATATION, DECYANIDATION, NEUTRALIZATION)
190201	metal hydroxide sludges and other sludges from metal insolubilization treatment
1904	VITRIFIED WASTES AND WASTES FROM VITRIFICATION
190402	fly ash and other flue gas treatment wastes
190403	non-vitrified solid phase
1908	WASTES FROM WASTE WATER TREATMENT PLANTS NOT OTHERWISE SPECIFIED
190803	grease and oil mixture from oil/waste water separation
190806	saturated or spent ion exchange resins
190807	solutions and sludges from regeneration of ion exchangers
20	MUNICIPAL WASTES AND SIMILAR COMMERCIAL, INDUSTRIAL AND INSTITUTIONAL WASTES INCLUDING SEPARATELY COLLECTED FRACTIONS
2001	SEPARATELY COLLECTED FRACTIONS
200112	paint, inks, adhesives and resins
200113	solvents
200117	photo chemicals
200119	pesticides
200121	fluorescent tubes and other mercury containing waste

Index